Tucholsky Wagner Zola Scott Sydow Freud Schlegel
Turgenev Fonatne Wallace
Twain Walther von der Vogelweide Fouqué Friedrich II. von Preußen
Weber Freiligrath Frey
Kant Ernst
Fechner Fichte Weiße Rose von Fallersleben Richthofen Frommel
Hölderlin
Engels Fielding Eichendorff Tacitus Dumas
Fehrs Faber Flaubert
Eliasberg Ebner Eschenbach
Feuerbach Maximilian I. von Habsburg Fock Zweig
Ewald Eliot Vergil
Goethe London
Elisabeth von Österreich
Mendelssohn Balzac Shakespeare Dostojewski Ganghofer
Lichtenberg Rathenau Doyle Gjellerup
Trackl Stevenson Hambruch
Mommsen Tolstoi Lenz Droste-Hülshoff
Thoma Hanrieder
von Arnim Hägele Humboldt
Dach Verne Hauff
Reuter Rousseau Hagen Hauptmann Gautier
Karrillon Garschin Baudelaire
Defoe Hebbel
Damaschke Descartes
Hegel Kussmaul Herder
Wolfram von Eschenbach Dickens Schopenhauer Rilke George
Darwin Melville Grimm Jerome
Bronner Bebel
Campe Horváth Aristoteles Proust
Voltaire Federer
Bismarck Vigny Barlach Herodot
Gengenbach Heine
Storm Casanova Tersteegen Grillparzer Georgy
Chamberlain Lessing Langbein Gilm
Gryphius
Brentano Lafontaine
Claudius Schiller Kralik Iffland Sokrates
Strachwitz Bellamy Schilling
Katharina II. von Rußland Gerstäcker Raabe Gibbon Tschechow
Löns Hesse Hoffmann Gogol Wilde Gleim Vulpius
Luther Heym Hofmannsthal Morgenstern
Roth Klee Hölty Goedicke
Heyse Klopstock Kleist
Luxemburg Puschkin Homer Mörike
La Roche Horaz Musil
Machiavelli Kierkegaard Kraft Kraus
Navarra Aurel Musset
Lamprecht Kind Kirchhoff Hugo Moltke
Nestroy Marie de France
Laotse Ipsen Liebknecht
Nietzsche Nansen
Marx Ringelnatz
von Ossietzky Lassalle Gorki Klett Leibniz
May vom Stein Lawrence Irving
Petalozzi Knigge
Platon Kafka
Sachs Pückler Michelangelo Kock
Poe Liebermann
de Sade Praetorius Mistral Zetkin Korolenko

The publishing house tredition has created the series **TREDITION CLASSICS**. It contains classical literature works from over two thousand years. Most of these titles have been out of print and off the bookstore shelves for decades.

The book series is intended to preserve the cultural legacy and to promote the timeless works of classical literature. As a reader of a **TREDITION CLASSICS** book, the reader supports the mission to save many of the amazing works of world literature from oblivion.

The symbol of **TREDITION CLASSICS** is Johannes Gutenberg (1400 – 1468), the inventor of movable type printing.

With the series, tredition intends to make thousands of international literature classics available in printed format again – worldwide.

All books are available at book retailers worldwide in paperback and in hardcover. For more information please visit: www.tredition.com

tredition was established in 2006 by Sandra Latusseck and Soenke Schulz. Based in Hamburg, Germany, tredition offers publishing solutions to authors and publishing houses, combined with worldwide distribution of printed and digital book content. tredition is uniquely positioned to enable authors and publishing houses to create books on their own terms and without conventional manufacturing risks.

For more information please visit: www.tredition.com

The Fern Lover's Companion A Guide for the Northeastern States and Canada

George Henry Tilton

Imprint

This book is part of the TREDITION CLASSICS series.

Author: George Henry Tilton
Cover design: toepferschumann, Berlin (Germany)

Publisher: tredition GmbH, Hamburg (Germany)
ISBN: 978-3-8495-1222-4

www.tredition.com
www.tredition.de

[Illustration: A Fern Lover]

The Fern Lover's Companion

A Guide for the Northeastern States and Canada

BY

GEORGE HENRY TILTON, A.M.

"This world's no blot for us Nor blank; it means intensely and it means good To find its meaning is my meat and drink."

DEDICATION

To Alice D. Clark, engraver of these illustrations, who has spared no pains to promote the artistic excellence of this work, and to encourage its progress, these pages are dedicated with the high regards of THE AUTHOR.

CONTENTS

LIST OF ILLUSTRATIONS

- A Fern Lover
- Prothallium Diagram
- Pinnate Frond
- Bipinnate Frond
- Pinnatifid Frond
- Spore Cases
- Linen Tester
- Curly Grass. *Schiz泰/i>*
- *Cinnamon Fern. Osmunda cinnamomea*
- *Sensitive Fern. Onoclea sensibilis*
- *Ostrich Fern. Onoclea Struthiopteris*
- *Interrupted Fern. Osmunda Claytoniana*
- *Climbing Fern. Lygodium*
- *Flowering Fern. Osmunda regalis spectabilis*
- *Adder's Tongue. Ophioglossum*
- *Grape Fern. Botrychium*
- *Polypody. Polypodium*
- *Beech Fern. Phegopteris*
- *Cloak Fern. Nothol殡*
- *Filmy Fern. Trichomanes*
- *Bracken. Pteris*
- *Maidenhair. Adiantum*
- *Cliff Brake. Pell泰/i>*
- *Lip Fern. Cheilanthes*
- *Rock Brake. Cryptogramma*
- *Chain Fern. Woodwardia*
- *Shield Fern. Polystichum*
- *Wood Fern. Aspidium*
- *Bladder Fern. Cystopteris*
- *Woodsia*

- *Osmunda cinnamomea glandulosa*
- *Curly Grass. Schiz[illegible]pusilla*
- *Sporangia of Curly Grass*
- *Climbing Fern. Lygodium palmatum*
- *Adder's Tongue. Ophioglossum vulgatum*
- *Moonwort. Botrychium Lunaria*
- *Moonwort, Details*
- *Little Grape Fern. Botrychium simplex*
- *Lance-leaved Grape Fern. Botrychium lanceolatum*
- *Matricary Grape Fern. Botrychium ramosum*
- *Common Grape Fern. Botrychium obliquum*
- *Botrychium obliquum var. dissectum*
- *Botrychium obliquum var. oneidense*
- *Ternate Grape Fern. Botrychium ternatum var. intermedium*
- *Ternate Grape Fern. B. ternatum var. intermedium*
- *Rattlesnake Fern. Botrychium virginianum*
- *Filmy Fern. Trichomanes Boschianum*
- *Fruiting Pinnules of Filmy Fern*
- *Crosiers*
- *Noted Fern Authors*
- *Spray of the Bulblet Bladder Fern*

PREFACE

A lover of nature feels the fascination of the ferns though he may know little of their names and habits. Beholding them in their native haunts, adorning the rugged cliffs, gracefully fringing the water-courses, or waving their stately fronds on the borders of woodlands, he feels their call to a closer acquaintance. Happy would he be to receive instruction from a living teacher: His next preference would be the companionship of a good fern book. Such a help we aim to give him in this manual. If he will con it diligently, consulting its glossary for the meaning of terms while he quickens his powers of observation by studying real specimens, he may hope to learn the names and chief qualities of our most common ferns in a single season.

Our most productive period in fern literature was between 1878, when Williamson published his "Ferns of Kentucky," and 1905, when Clute issued, "Our Ferns in Their Haunts." Between these flourished D.C. Eaton, Davenport, Waters, Dodge, Parsons, Eastman, Underwood, A.A. Eaton, Slosson, and others. All their works are now out of print except Clute's just mentioned and Mrs. Parsons' "How to Know the Ferns." Both of these are valuable handbooks and amply illustrated. Clute's is larger, more scholarly, and more inclusive of rare species, with an illustrated key to the genera; while Mrs. Parsons' is more simple and popular, with a naive charm that creates for it a constant demand.

We trust there is room also for this unpretentious, but progressive, handbook, designed to stimulate interest in the ferns and to aid the average student in learning their names and meaning. Its geographical limits include the northeastern states and Canada. Its nomenclature follows in the main the seventh edition of Gray's Manual, while the emendations set forth in Rhodora, of October, 1919, and also a few terms of later adoption are embodied, either as synonyms or substitutes for the more familiar Latin names of the Manual, and are indicated by a different type. In every case the student has before him both the older and the more recent terms from which to choose. However, since the book is written primarily for lovers of Nature, many of whom are unfamiliar with scientific terms, the common English names are everywhere given prominence, and strange to say are less subject to change and controversy than the Latin. There is no doubt what species is meant when one speaks of the Christmas fern, the ostrich fern, the long beech fern, the interrupted fern, etc. The use of the common names will lead to the knowledge and enjoyment of the scientific terms.

A friend unfamiliar with Latin has asked for pointers to aid in pronouncing the scientific names of ferns. Following Gray, Wood, and others we have marked each accented syllable with either the grave (`) or acute () accent, the former showing that the vowel over which it stands has its long sound, while the latter indicates the short or modified sound. Let it be remembered that any syllable with either of these marks over it is the accented syllable, whose sound will be long or short according to the slant of the mark.

We have appropriated from many sources such material as suited our purpose. Our interest in ferns dates back to our college days at Amherst, when we collected our first specimens in a rough, bushy swamp in Hadley. We found here a fine colony of the climbing fern (Lygodium). We recall the slender fronds climbing over the low bushes, unique twiners, charming, indeed, in their native habitat. We have since collected and studied specimens of nearly every New England fern, and have carefully examined most of the other species mentioned in this book. By courtesy of the librarian, Mr. William P. Rich, we have made large use of the famous Davenport herbarium in the Massachusetts Horticultural library, and through the kindness of the daughter, Miss Mary E. Davenport, we have freely consulted the larger unmounted collection of ferns at the Davenport homestead, at Medford,[1] finding here a very large and fine assortment of Botrychiums, including a real B. ternatum from Japan.

[Footnote 1: Recently donated to the Gray Herbarium.]

For numerous facts and suggestions we are indebted to the twenty volumes of the Fern Bulletin, and also to its able editor, Mr. Willard N. Clute. To him we are greatly obligated for the use of photographs and plates, and especially for helpful counsel on many items. We appreciate the helpfulness of the American Fern Journal and its obliging editor, Mr. E.J. Winslow. To our friend, Mr. C.H. Knowlton, our thanks are due for the revision of the checklist and for much helpful advice, and we are grateful to Mr. S.N.F. Sanford, of the Boston Society of Natural History, for numerous courtesies;

but more especially to Mr. C.A. Weatherby for his expert and helpful inspection of the entire manuscript.

The illustrations have been carefully selected; many of them from original negatives bequeathed to the author by his friend, Henry Lincoln Clapp, pioneer and chief promoter of school gardens in America. Some have been photographed from the author's herbarium, and from living ferns. A few are from the choice herbarium of Mr. George E. Davenport, and also a few reprints have been made from fern books, for which due credit is given. The Scott's spleenwort, on the dedication page, is reprinted from Clute's "Our Ferns in Their Haunts."

INTRODUCTION

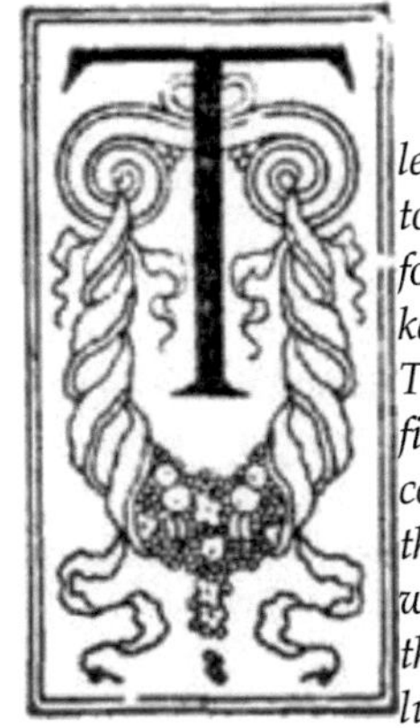

Thoreau tells us, "Nature made a fern for pure leaves." Fern leaves are in the highest order of cryptogams. Like those of flowering plants they are reinforced by woody fibres running through their stems, keeping them erect while permitting graceful curves. Their exquisite symmetry of form, their frequent finely cut borders, and their rich shades of green combine to make them objects of rare beauty; while their unique vernation and method of fruiting along with their wonderful mystery of reproduction invest them with marked scientific interest affording stimulus and culture to the thoughtful mind. By peculiar enchantments these charming plants allure the ardent Nature-lover to observe their haunts and habits.

"Oh, then most gracefully they wave In the forest, like a sea, And dear as they are beautiful Are these fern leaves to me."

As a rule the larger and coarser ferns grow in moist, shady situations, as swamps, ravines, and damp woods; while the smaller ones are more apt to be found along mountain ranges in some dry and even exposed locality. A tiny crevice in some high cliff is not infrequently chosen by these fascinating little plants, which protect themselves from drought by assuming a mantle of light wool, or of hair and chaff, with, perhaps, a covering of white powder as in some cloak ferns--thus keeping a layer of moist air next to the surface of the leaf, and checking transpiration.

Some of the rock-loving ferns in dry places are known as "resurrection" ferns, reviving after their leaves have turned sere and brown. A touch of rain, and lo! they are green and flourishing.

Ferns vary in height from the diminutive filmy fern of less than an inch to the vast tree ferns of the tropics, reaching a height of sixty feet or more.

REPRODUCTION

Ferns are propagated in various ways. A frequent method is by perennial rootstocks, which often creep beneath the surface, sending up, it may be, single fronds, as in the common bracken, or graceful leaf-crowns, as in the cinnamon fern. The bladder fern is propagated in part from its bulblets, while the walking leaf bends over to the earth and roots at the tip.

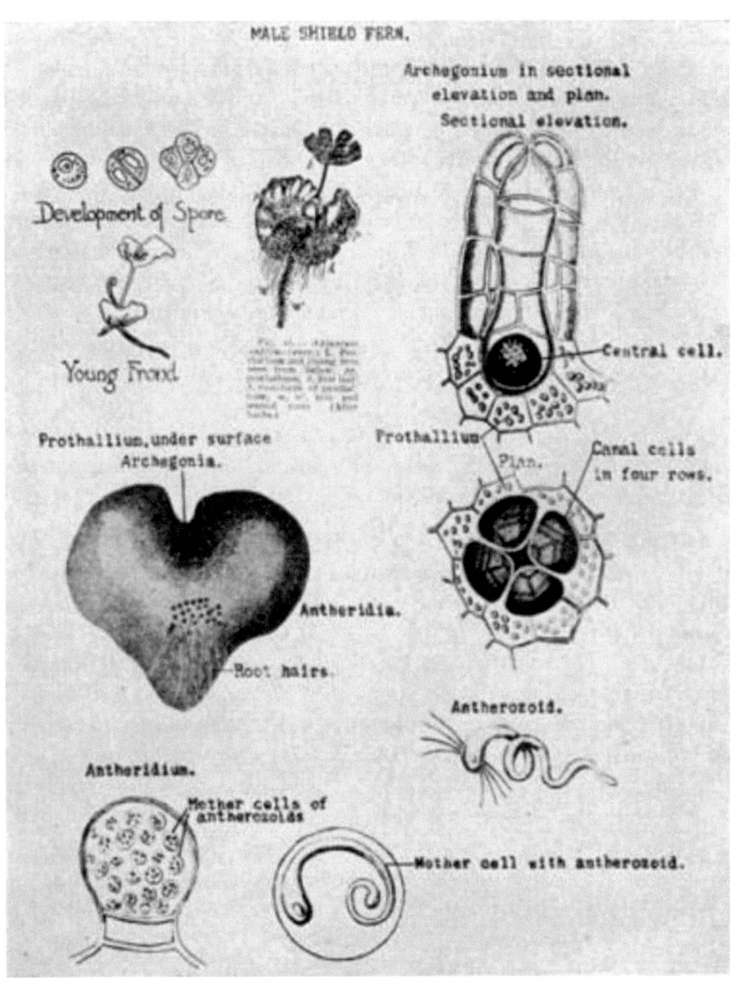

[Illustration: MALE SHIELD FERN. Fern Reproduction by the Prothallium]

Ferns are also reproduced by spores, a process mysterious and marvellous as a fairy tale. Instead of seeds the fern produces spores, which are little one-celled bodies without an embryo and may be likened to buds. A spore falls upon damp soil and germinates, producing a small, green, shield-shaped patch much smaller than a dime, which is called a proth • ium (or prothallus). On its under surface delicate root hairs grow to give it stability and nutriment; also two sorts of reproductive organs known as anther□a and archeg□, the male and female growths analogous to the stamens and pistils in flowers. From the former spring small, active, spiral bodies called • heroz•, which lash about in the moisture of the proth • ium until they find the archeg • , the cells of which are so arranged in each case as to form a tube around the central cell, which is called the • here, or egg-cell, the point to be fertilized. When one of the entering • heroz• reaches this point the desired change is effected, and the canal of the archeg • m closes. The empty • here becomes the quickened • hore whose newly begotten plant germ unfolds normally by the multiplication of cells that become, in turn, root, stem, first leaf, etc., while the proth • ium no longer needed to sustain its offspring withers away.[1]

[Footnote 1: In the accompanying illustration, it should be remembered that the reproductive parts of a fern are microscopic and cannot be seen by the naked eye.]

Fern plants have been known to spring directly from the proth • us by a budding process apart from the organs of fertilization, showing that Nature "fulfills herself in many ways."[2]

[Footnote 2: The scientific term for this method of reproduction is ap • y (apart from marriage). Sometimes the prothallus itself buds directly from the frond without spores, for which process the term ap • ry is used. (Meaning, literally, without spores.)]

VERNATION

All true ferns come out of the ground head foremost, coiled up like a watch-spring, and are designated as "fiddle-heads," or crosiers. (A real crosier is a bishop's staff.) Some of these odd young growths are covered with "fern wool," which birds often use in lining their nests. This wool usually disappears later as the crosier unfolds into the broad green blade. The development of plant shoots from the bud is called vernation (Latin, ver meaning spring), and this unique uncoiling of ferns, "circinnate vernation."

VEINS

The veins of a fern are free, when, branching from the mid-vein, they do not connect with each other, and simple when they do not fork. When the veins intersect they are said to anastomose (Greek, an opening, or network), and their meshes are called ar诬栯r •oles (Latin, areola, a little open space).

EXPLANATION OF TERMS

A frond is said to be pinnate (Latin, *pinna,* a feather), when its primary divisions extend to the rachis, as in the Christmas fern (Fig. 1). A frond is bipinnate (Latin, *bis,* twice) when the lobes of the pinnæ extend to the midvein as in the royal fern (Fig. 2). These divisions of the pinnæ are called pinnules. When a frond is trip·ate the last complete divisions are called ultimate pinnules or segments. A frond is pinnatifid when its lobes extend halfway or more to the rachis or midvein as in the middle lobes of the pinnatifid spleenwort (Fig. 3). The pinnæ of a frond are often pinnatifid when the frond itself is pinnate; and a frond may be pinnate in its lower part and become pinnatifid higher up as in the pinnatifid spleenwort just mentioned (Fig. 3).

The divisions of a pinnatifid leaf are called segments; of a bipinnatifid or tripinnatifid leaf, ultimate segments.

[Illustration: Fig. 1]

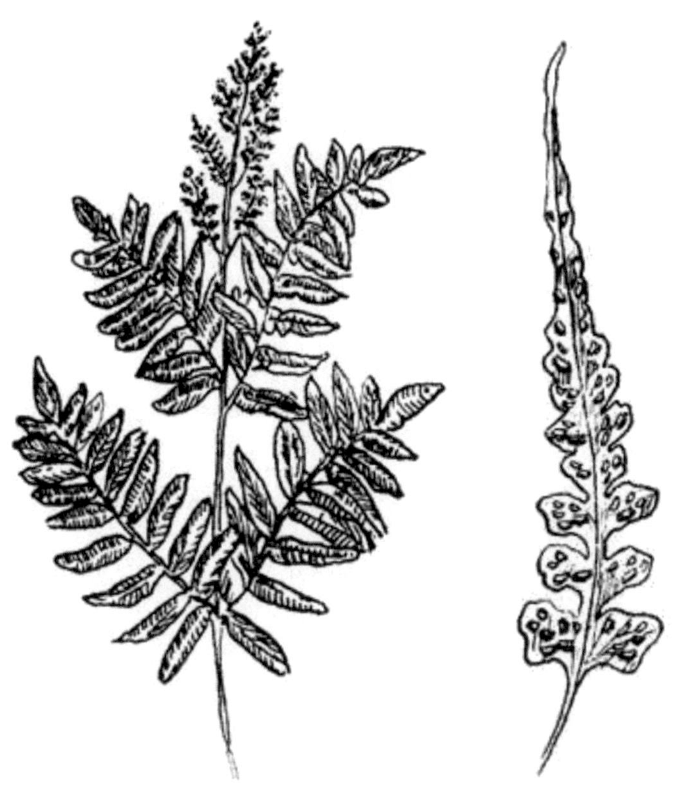

[Illustration: Fig. 2]   [Illustration: Fig. 3]

[Illustration: Fig. 4]

SPOR

GIA AND FRUIT DOTS

Fern spores are formed in little sacs known as spore-cases or spor•ia (Fig. 4). They are usually clustered in dots or lines on the back or margin of a frond, either on or at the end of a small vein, or in spike-like racemes on separate stalks. Sori (singular sorus, a heap), or fruit dots may be naked as in the polypody, but are usually covered with a thin, delicate membrane, known as the indusium (Greek, a dress, or mantle). The family or genus of a fern is often determined by the shape of its indusium; e.g., the indusium of the woodsias is star-shaped; of the Dicksonias, cup-shaped; of the aspleniums, linear; of the wood ferns, kidney-shaped, etc.

In many ferns the sporangia are surrounded in whole or in part by a vertical, elastic ring (annulus) reminding one of a small, brown worm closely coiled (Fig. 4). As the spores mature, the ring contracts and bursts with considerable force, scattering the spores. The spores of the different genera mature at different times from May to September. A good time to collect ferns is just before the fruiting season. (For times of fruiting see individual descriptions or chronological chart on page 220.)

HELPFUL HINTS

The following hints may be helpful to the young collector:

1. A good lens with needles for dissecting is very helpful in examining the sori, veins, glands, etc., as an accurate knowledge of any one of these items may aid in identifying a given specimen. Bausch and Lomb make a convenient two-bladed

pocket glass for about two dollars.[1]

[Footnote 1: In the linen tester here figured (cost $1.50) the lens is mounted in a brass frame which holds it in position, enabling the dissector to use both hands. A tripod lens will also be found cheap and serviceable.]

[Illustration: Linen Tester]

2. Do not exterminate or weaken a fern colony by taking more plants than it can spare. In small colonies of rare ferns take a few and leave the rest to grow. It is decidedly ill-bred to rob a locality of its precious plants. Pick your fern leaf down close to the root-stock, including a portion of that also, if it can be spared. Place your fronds between newspaper sheets and lay "dryers" over them (blotting paper or other absorbent paper). Cover with a board or slat frame, and lay on this a weight of several pounds, leaving it for twenty-four hours; if the specimens are not then cured, change the dryers. Mount the prepared specimens on white mounting sheets. The regulation size is 16-1/2 by 11-1/2 inches. The labels are usually 3-3/4 by 1-3/4 inches. A sample will suggest the proper inscription.

> *HERBARIUM OF JOHN DOE Ophioglm vulgatum, L. (Adder's Tongue) Willoughby Lake, Vt. August 19, 1911. Wet meadow. Coll. X.Y.Z. Rather common but often overlooked*

Place the label at the lower right-hand corner of the sheet, which is now ready to be laid in the genus cover, usually of manila paper 16-1/2 by 12 inches. It is well to jot down important memoranda at the time of collecting. This is the method in use at the Gray Herbarium in Cambridge. It can, of course, be modified to suit one's own taste or convenience. The young collector can begin by simply pressing his specimens between the leaves of a book, the older and coarser the better; and he can mount them in a blank book designed for the purpose, or if he has only a common blank book, he can cut out some of the leaves, alternately with others left in place, as is often done with a scrap book, that when the book is full it may not be crowded at the back. Or he can use sheets of blank paper of any uniform size and mount the specimens on these with gummed strips, and then group them, placing those of the same genus together. Such an extemporized herbarium, though crude, will serve for a beginning, while stimulating his interest, and advancing his knowledge of the ferns. Let him collect, press, and mount as many varieties as possible, giving the name with date and place of collecting, etc. Such a first attempt may be kept as a reminder of pleasant hours spent in learning the rudiments of a delightful study.

We cannot insist too strongly upon the necessity of handling and studying the living plant. Every student needs to observe for himself the haunts, habits, and structure of real ferns. We would say to the young student, while familiarizing yourself with the English names of the ferns, do not neglect the scientific names, which often hold the key to their meaning. Repeat over and over the name of each genus in soliloquy and in conversation until your mind instantly associates each fern with its family name-- "Adiantum," "Polystichum," "Asplenium," and all the rest. Fix them in the memory for a permanent asset. With hard study and growing knowledge will come growing attachment. How our great expert, Mr. Davenport, loved the ferns! He would handle them with gentle touch, fondly stroke their leaves, and devoutly study their structure, as if inspired by the All-wise Interpreter.

> *"Move along these shades In gentleness of heart: with gentle hand Touch--for there is a spirit in the woods."*

KEY TO THE GENERA

This key, in illustrating each genus, follows the method of Clute in "Our Ferns in Their Haunts," but substitutes other and larger specimens. Five of these are from Waters' "Ferns" by permission of Henry Holt & Co.

As the indusium, which often determines the name of a fern, is apt in some species to wither early, it is important to secure for study not only a fertile frond, but one in as good condition as possible. For convenience the ferns may be considered in two classes.

I

THOSE WHICH HAVE THE FRUITING PORTION IN GREENISH, BERRY-LIKE STRUCTURES AND NOT ON THE BACK OF FRONDS

A. FRUITING FRONDS WHOLLY FERTILE
(Fertile and sterile fronds entirely unlike)

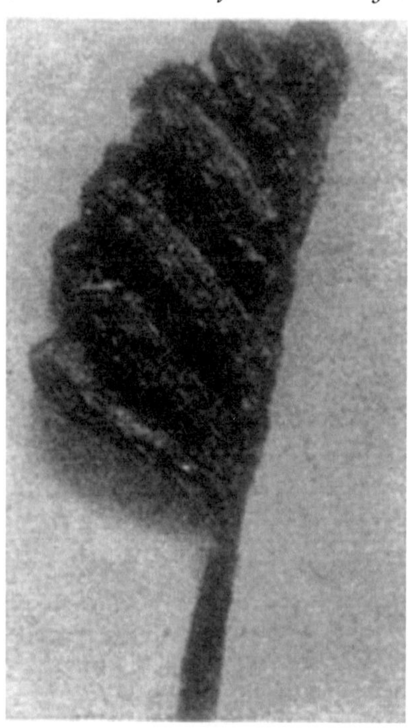

1. Fruit in a one-sided spike in two ranks; plants very small; sterile fronds thread-like and tortuous.

Curly Grass. *Schiz泰/i>.*

2. Fruit in a club-shaped, brown or cinnamon-colored spike loaded with sporangia; fruit in early spring.

Cinnamon Fern. *Osmunda cinnamomea*.

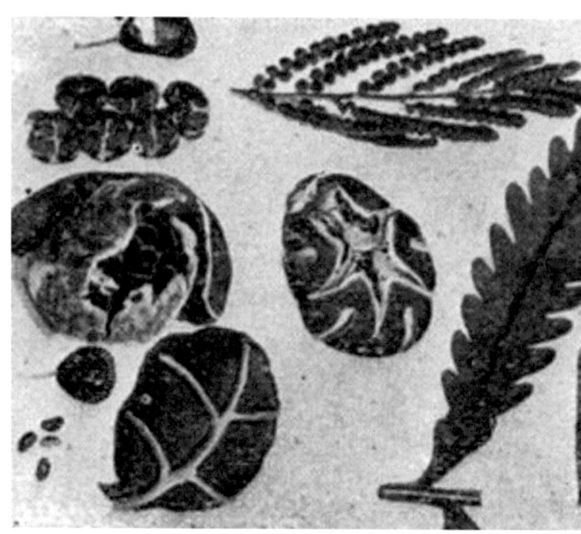

3. Fruit in berry-like, greenish structures in a twice pinnate spike, which comes up much later than the broad and coarse pinnɐfid sterile fronds.

Wet ground. Sensitive Fern. *Onoclea.*

4. Fruit in pod-like or neck-
lace-like pinn漌fertile frond
pinnate; sterile frond tall,
pinnᴇfid; fruit late.

Ostrich Fern. *Onoclea struthiopteris.*

B. FRUITING FRONDS PARTLY STERILE

1. Fruiting portion in the middle of the frond; two to four pairs of fertile pinn殻/p>

Interrupted Fern. *Osmunda Claytoniana.*

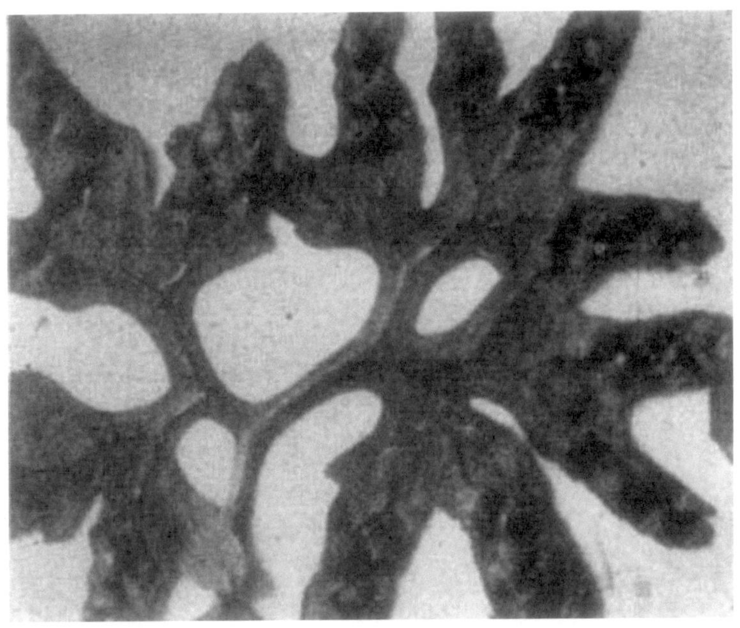

2. Fruiting portion at the apex of the frond. Sterile pinn枕almate; rachis twining.

Climbing Fern. *Lygodium.*

Sterile pinn枙innate; fronds large, fertile portion green, turning brown, forming a panicle at the top.

Royal Fern. *Osmunda regalis.*

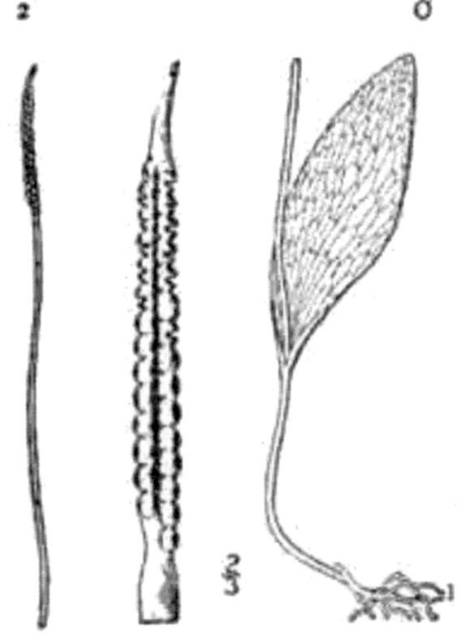

3. Fruiting portion seemingly on a separate stock a few inches above the sterile.

Sterile part an entire, ovate, green leaf near the middle; fertile part a spike.

Adder's Tongue. *Ophioglossum*.

Sterile portion more or less divided; fruit in racemes or panicles, rarely in spikes.

Grape Ferns. Moonwort. *Botrychium*.

II

THOSE WHICH HAVE THE FRUITING PORTION ON THE BACK OR MARGIN OF FRONDS

A. INDUSIUM WANTING

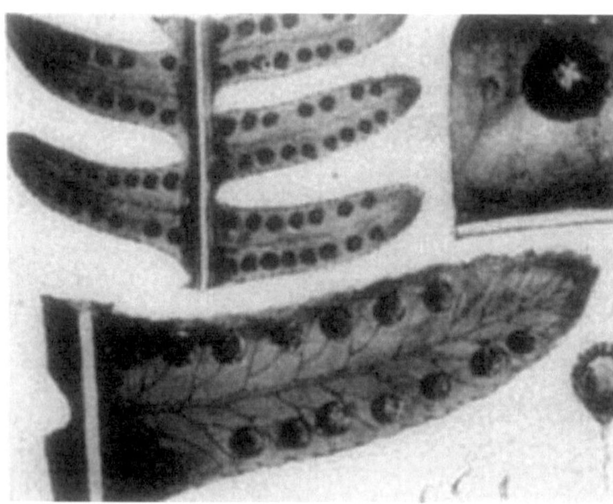

1. Fruit-dots large, roundish; fronds evergreen. Rock species.

Polypody. *Polypodium.*

2. Fruit-dots small, roundish; fronds triangular.

Beech Ferns. *Phegopteris.*

3. Fruit in lines on the margin of the pinnules; under surface of the fronds covered with whitish powder.

Cloak Ferns. *Nothol殯.*

B. INDUSIUM PRESENT

1. Sori on the edge of a pinnule terminating a vein; sporangia at the base of a long, bristle-like receptacle surrounded by a cup-shaped indusium.

Filmy Fern. *Trichomanes*.

2. Indusium formed by the reflexed margin of the pinnules.

(1) Sporangia on a continuous line; fronds large, ternate; indusium narrow. Bracken. Brake. *Pteris*.

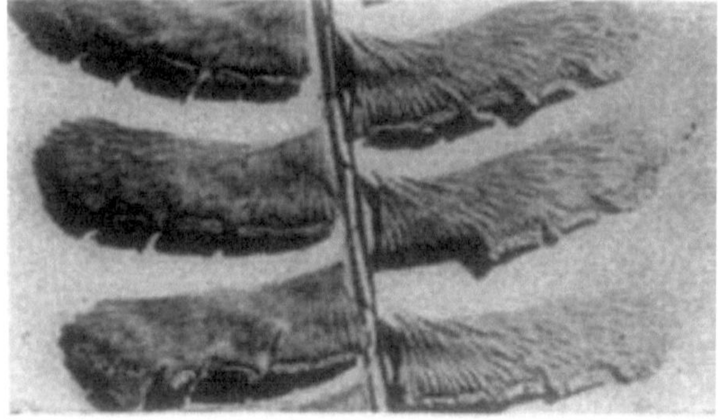

(2) Sporangia in oblong sori under a reflexed tooth of a pinnule; indusium broad; rachis dark and shining. Maidenhair.

(3) Sori in roundish or elongated masses.

Indusium broad, nearly continuous, fronds mostly smooth, somewhat leathery, pinnate. Rock species. Cliff brakes. *Pell泰/i>*.

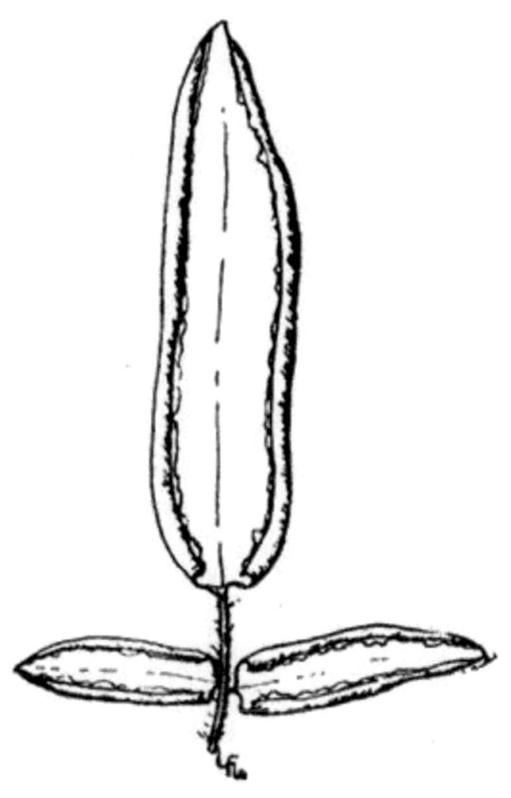

Indusium narrow, seldom continuous, formed by the margin of separate lobes or of the whole pinnules; often inconspicuous, fronds usually hairy. Lip Ferns. *Cheilanthes*.

Indusium of the reflexed edges, at first reaching to the midrib, or nearly so; later opening out nearly flat; fruiting pinnules pod-like; sterile fronds broad. Rock brakes. *Cryptogramma*.

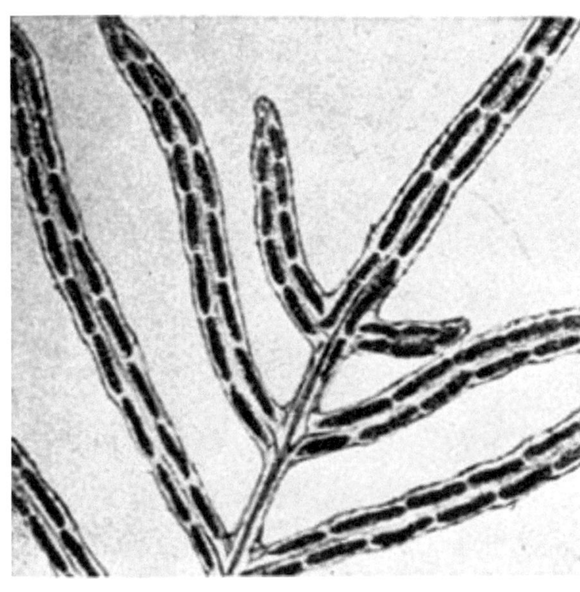

3. Indusium never formed of the margin of the frond. Sori various.

(1) Fruit-dots oblong, parallel with the midrib, somewhat sunken in the tissues of the frond. Water-loving species. Chain Ferns. *Woodwardia.*

(2) Fruit-dots and indusium roundish.

Indusium shield-shaped, fixed by the center. Evergreen glossy ferns in rocky woods. Shield Ferns. *Polystichum.*

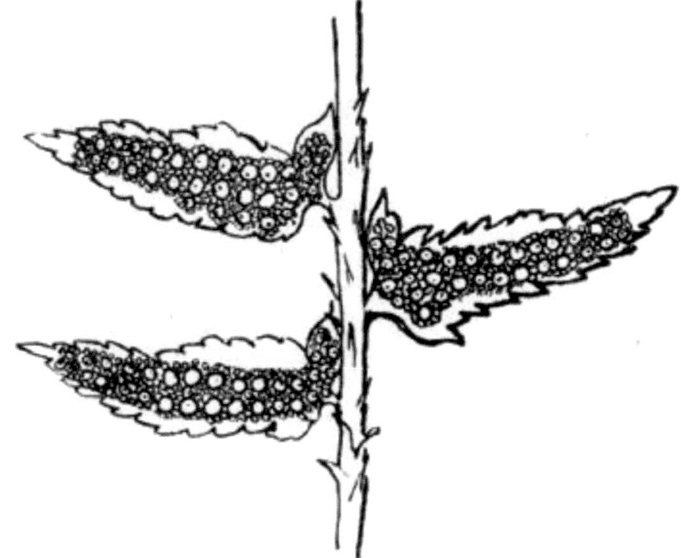

Indu-
sium
cordate,
fixed by
the sinus.
Wood
Ferns.
Aspidium.

Indusium hood-shaped, fixed centrally behind the sorus and arching over it, soon withering, often illusive. Fronds two to three pinnate, very graceful. Moisture-loving species. Bladder Ferns. *Cystopteris.*

Indusium star-shaped, of a few irregular segments fixed beneath the sorus, often obscure. Mostly small, rock-loving plants, usually rather chaffy, at least at the base, and growing in tufts. *Woodsia.*

Indusium cup-shaped, fixed beneath the sorus, supported by the tooth of a leaf; sporangia borne in an elevated, globular receptacle open at the top. Fronds finely cut. Hayscented Fern. *Dennst椴ia.*

(3) Fruit-dots and indusium linear. (But see *Athyrium*.)

Very long, nearly at right angles to the midrib, double; blade thick oblong-lanceolate, entire; heart-shaped at the base. Hart's Tongue. *Scolopendrium.*

Shorter and irregularly scattered on the under side of the frond, some parallel to the midrib, others oblique to it, and often in pairs or joined at the ends; blade tapering to a slender tip. Walking Fern. *Campto-sorus.*

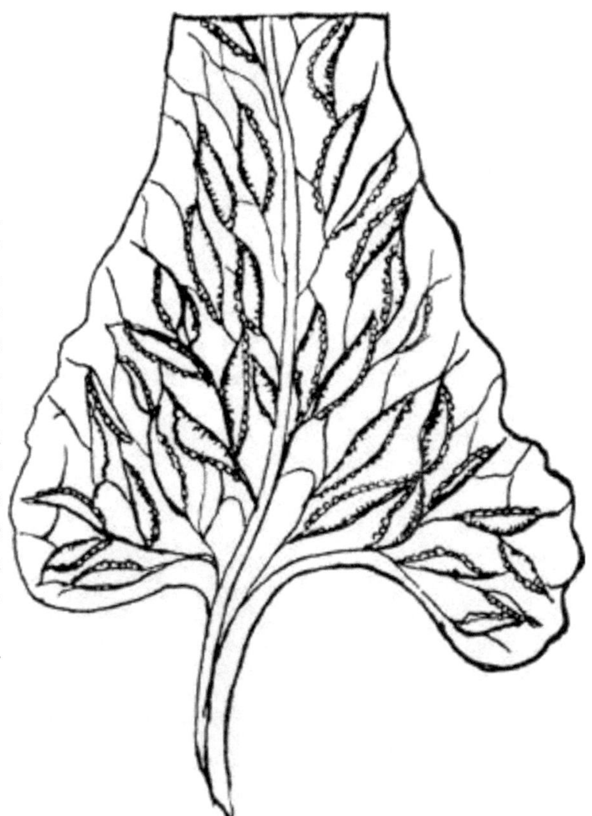

Short,
straight,
mostly
oblique
the mi
Indusiu
rather
row, o
ing to
the mi
fronds l
or vario
divided
Spleenw
Aspleniu

Short, in-
dusium
usually
more or less
curved and
frequently
crossing a
vein. The
large sple-
enworts
including
Lady Fern.
Athyrium.

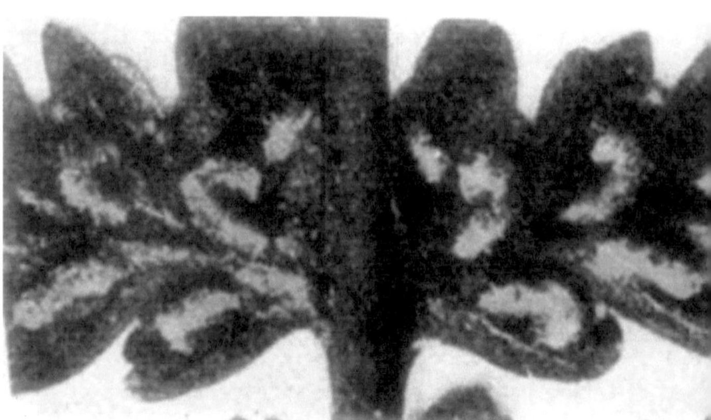

DESCRIPTIVE TEXT OF THE FERNS

In this manual our native ferns are grouped scientifically under five distinct families. By far the largest of these groups, and the first to be treated, is that of the real ferns (Polypodi□檪/i> with sixty species and several chief varieties. Then follow the flowering ferns (Osmund ·檪/i> with three species; the curly grass and climbing ferns (Schiz栬e檪/i> with two species; the adder's tongue and grape ferns (Ophiogloss ·檪/i> with seven species; and the filmy ferns (Hymenophyll ·檪/i> with one species.

Corresponding with these five families, the sporangia or spore cases of ferns have five quite distinct forms on which the families are founded.

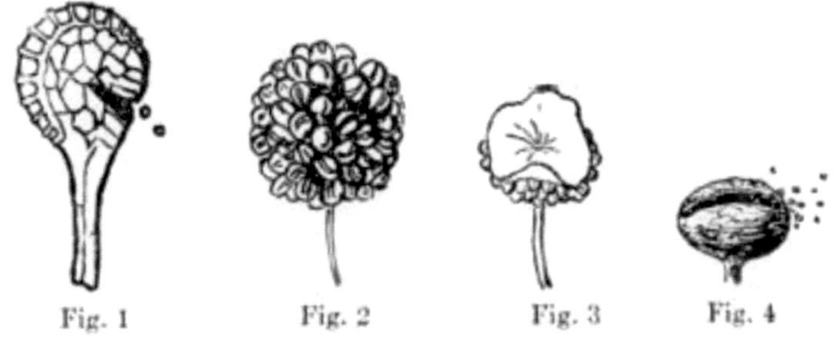

Fig. 1-4: Sporangia of the Five Families]

1. The Fern Family proper (Polypodi ·潹i>) has the spore cases stalked and bound by a vertical, elastic ring (Fig. 1). The clusters of fruit-dots containing the spore cases may be open and naked as in polypody (Fig. 2), or covered by an indusium, as in the shield ferns (Fig. 3).

2. The Royal Fern Family (Osmunda) has the spore cases stalked with only a rudimentary ring on one side, which opens longitudinally (Fig. 4).

3. The Climbing Fern Family (Lygodium, Schiz泰/i>) has the spore cases sessile in rows; they are small, nut-like bodies with the elastic ring around the upper portion (Fig. 5).[1]

[Footnote 1: These figures are enlarged.]

4. The Adder's Tongue Family (Ophiogl ·m, Botrchium) has simple spore cases without a ring, and discharges its spores through a transverse slit (Fig. 6).

5. The Filmy Fern Family (Trich□es) has the spore cases along a bristle-like receptacle and surrounded by an urn-shaped, slightly two-lipped involucre; ring transverse and opening vertically (Fig. 7).

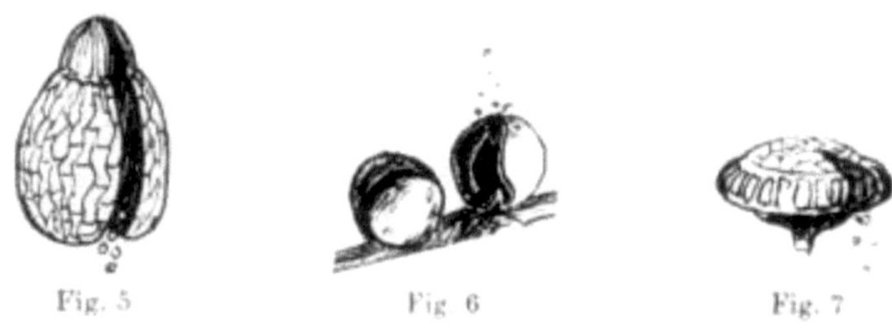

Fig. 5-7: Indusium

THE FERN FAMILY PROPER OR REAL FERNS

POLYPODI ᴸE5/i>

Green, leafy plants whose spores are borne in spore-cases (sporangia), which are collected in dots or clusters (fruit-dots or sori) on the back of the frond or form lines along the edge of its divisions. Sporangia surrounded by vertical, elastic rings bursting transversely and scattering the spores. Fruit-dots (sori) often covered, at least when young, by a membrane called the indusium. Spores brown.

THE POLYPODIES

1. POLYPODY. Polypodium

(From the Greek meaning many-footed, alluding to the branching rootstocks.)

Simple ferns with stipes articulated to the creeping rootstocks, which are covered with brown, chaffy scales. Fruit-dots round, naked, arranged on the back of the frond in one or more rows each side of the midrib. Sporangia pedicelled, provided with a vertical ring which bursts transversely. A large

*genus with about 350 species, widely distributed, mostly in tropical re-
gions.*

(1) COMMON POLYPODY. *Polypodium vulgare*

*Fronds somewhat leathery in texture, evergreen, four to ten inches tall,
smooth, oblong, and nearly pinnate. The large fruit-dots nearly midway
between the midrib and the margin, but nearer the margin.*

Common Polypody. Polypodium vulgare

Common everywhere on cliffs, usually in half shade, and may at times spring out of decaying logs or the trunks of trees. As the jointed stipes, harking back to some ancient mode of fern growth, fall away from the rootstocks after their year of greenness, they leave behind a scar as in Solomon's seal. The polypody is a gregarious plant. By intertwining its roots the fronds cling together in "cheerful community," and a friendly eye discovers their beauty a long way off. August. Abounds in every clime, including Europe and Japan.

In transplanting, sections should be cut, not pulled from the matted mass.

Var. *cambricum* has segments broader and more or less strongly toothed.

Var. *cristatum* has the segments forked at the ends.

Several other forms are also found.

Fruited Frond

The Common Polypody. *Polypodium vulgare* (Photographed by Miles Greenwood, Melrose, Mass.)

(2) GRAY OR HOARY POLYPODY

Polypodium incவௗm. P. polypodi□s

Fronds oblong, two to seven inches long, deeply pinnρfid, gray and scurfy underneath with peltate scales having a dark center. Fruit-dots rather small, near the margin and obscured by the chaff.

Gray or Hoary Polypody. *Polypodium incanum*

In appearance the gray polypody is much like the common species, as the Greek ending oides (like) implies. In Florida and neighboring states it often grows on trees; farther north mostly on rocks. Reported as far north as Staten Island. It is one of the "resurrection" ferns, reviving quickly by moisture after seeming to be dead from long drouth. July to September. Widely distributed in tropical America. Often called Tree-Polypody.

THE BRACKEN GROUP

Sporangia near or on the margin of the segments, the reflexed portions of which serve as indusia.

1. BRACKEN OR BRAKE

Pt版s aquilina. PTERIUM LATI₂CULUM[1]

[Footnote 1: The use of small capitals in the scientific names indicates in part the newer nomenclature which many botanists are inclined to adopt.]

Fronds broadly triangular, ternate, one to three feet high or more, the widely spreading branches twice pinnate, the lower pinnules more or less pinnᵖfid. Sporangia borne in a continuous line along the lower margin of the ultimate divisions whose reflexed edges form the indusium. (Greek, pteron, a wing, the feathery fronds suggesting the wings of a bird.)

Common Bracken or Brake, a Sterile Frond. Pteris aquilina (Providence County, R.I.)

A Fertile Frond of Common Bracken. Pteris aquilina (Suffolk County, Mass.)

"The heath this night must be my bed, The bracken curtain for my head." SCOTT.

The outlines of the young bracken resemble the little oak fern. It flourishes in thickets and open pastures, often with poor soil and scant shade. It *is found in all parts of the world, and is said to be the most common of all our North American ferns. In a cross section of the mature stipe superstition sees "the devil's hoof" and "King Charles in the oak," and any one may see or think he sees the outlines of an oak tree. It was the bracken, or eagle fern, as some call it, which was supposed to bear the mysterious "fern seed," but only on midsummer eve (St. John's eve).*

> *"But on St. John's mysterious night, Confest the mystic fern seed fell."*

This enabled its possessor to walk invisible.

> *"We have the receipt for fern-seed, We walk invisible." SHAKE-SPEARE.*

The word brake or bracken is one of the many plant names from which some of our English surnames are derived, as Brack, Breck, Brackenridge, etc., and fern (meaning the bracken) is seen in Fern, Fearns, Fernham, Fernel, Fernside, Farnsworth, etc. Also, in names of places as Ferney, Ferndale, Fernwood, and others. Although the bracken is coarse and common, it makes a desirable background for rockeries, or other fern masses. The young ferns should be transplanted in early spring with as much of the long, running rootstock as possible.

Var. pseudocaudata has longer, narrower and more distant pinnules, and is a common southern form.

Var. pseudocaudata

2. MAIDENHAIR. *Adiantum*

Ferns with much divided leaves and short, marginal sori borne at the ends of free-forking veins, on the under side of the reflexed and altered portion of the pinnules, which serves as an indusium. Stipes and branches of the leaves very slender and polished.

(Greek, unwetted, because drops of water roll off without wetting the leaves.)

(1) COMMON MAIDENHAIR. *Adiantum pedatum*

A graceful fern of shady glen and rocky woodland, nine to eighteen inches high, the black, shining stalks forked at the top into two equal, recurved branches, the pinn校ll springing from the upper side. Pinnules triangular-oblong, bearing short sori on their inwardly reflexed margins which form the indusium.

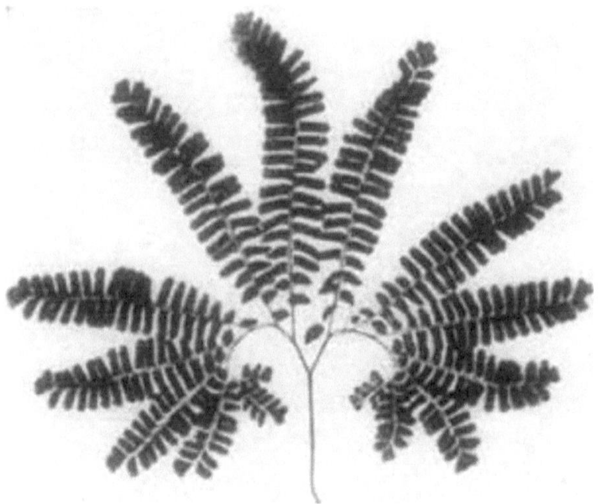

[Illustration: A Spray of Maidenhair]

[Illustration: Fruiting Pinn桷f Maidenhair]

The maidenhair has a superficial resemblance to the meadow rue, which also sheds water, but it may be known at once by its black, shining stalks with their divisions all borne on one side. It is indeed a most delicate fern, known and admired by every one. The term maidenhair may have been suggested by the black, wiry roots growing from the slender rootstock, or by the dark, polished stems, or, as Clute explains it, "because the black roots, like hair, were supposed, according to the 'doctrine of signatures' to be good for falling hair, and the plant was actually used in the 'syrup of capillaire'[A] (Am. Botanist, November, 1921). While the maidenhair is not very common, it is widely distributed, being found throughout our section, westward to California, and northward to the British Provinces.

"Though the maidenhair has a wide range, and grows abundantly in many localities, it possesses a quality of aloofness which adds to its charm. Its

chosen haunts are dim, moist hollows in the woods, or shaded hillsides sloping to the river. In such retreats you find the feathery fronds tremulous on their glistening stalks, and in their neighborhood you find, also, the very spirit of the woods."
MRS. PARSONS.

[Footnote A: It may be stated that capillaire syrup besides the use here indicated was highly esteemed as a pectoral for the relief of difficult breathing.]

[Illustration: Common Maidenhair. Adiantum pedatum (Reading, Mass., Kingman)]

[Illustration: Alpine Maidenhair. *Adiantum pedatum, Var. aleuticum* (Fernald and Collins, Gaspé County, Quebec, 1906) (From the Gray Herbarium)]

The fern is not hard to cultivate if allowed sufficient moisture and shade. Along with the ostrich fern it makes a most excellent combination in a fern border.

Var. ALEUTICUM, or Alpine Maidenhair. A beautiful northern form especially abundant on the high tableland of the Gasp預eninsula, Quebec, where it is said to cover hundreds of acres. In the east it is often dwarfed-- six to ten inches high, growing in tufts with stout rootstocks, having the pinnules finely toothed instead of rounded and the indusia often lunate, rarely twice as long as broad. (Fernald in Rhodora, November, 1905.) Also found in northern Vermont, and to the northwestward.

(2) THE VENUS-HAIR FERN. *Adiantum Cap •us-Veneris*

Fronds with a continuous main rachis, ovate-lanceolate, twice pinnate below. Pinnules, fan-shaped on slender, black stalks, long, deeply and ir- regularly incised. Veins extending from the base of the pinnules like the ribs of a fan.

[Illustration: Venus Hair Fern. Adiantum Capillus-Veneris]

While our common maidenhair is a northern fern, the Venus-hair Fern is confined to the southern states. It is rarely found as far north as Virginia, where it meets, but scarcely overlaps its sister fern. The medicinal properties of Adiantum pedatum were earlier ascribed to the more southern species, which is common in Great Britain, but, like many another old remedy, "the syrup of capillaire" is long since defunct.

3. CLIFF BRAKES. *Pellæa*

Sporangia borne on the upper part of the free veins inside the margins, in dot-like masses, but may run together, as in the continuous fruiting line of the bracken. Indusium formed of the reflexed margins of the fertile seg-

ments which are more or less membranous. (Pellæa from the Greek pellos, meaning dusky, in allusion to the dark stipes.)

(1) PURPLE CLIFF BRAKE. *Pellæa atropurprea*

Stipes dark purple or reddish-brown, polished and decidedly hairy and harsh to the touch, at least on one side. Fronds coriaceous, pale, simply pinnate, or bipinnate below; the divisions broadly linear or oblong, or the sterile sometimes oval, chiefly entire, somewhat heart-shaped, or else truncate at the stalked base. Veins about twice forked. Basal scales extending into long, slender tips, colorless or yellow.

[Illustration: Purple Cliff Brake. Pell枈atropurpurea]

Another name is "the winter brake," as its fronds remain green throughout the winter, especially in its more southern ranges. It grows on rocky ledges with a preference for limestone, and often in full sun. In large and mature fronds its pinn挣 are apt to be extremely irregular. While its stipes are purplish, its leaves are bluish-green, and its scales light-brown or yellow. Strange to say, this brake of the cliffs thrives in cultivation. Wool-

son says of it, "This fern is interesting and valuable. It is not only beautiful in design, but unique in color, a dark blue-green emphasizing all the varying tints about it--a first-class fern for indoor winter cultivation. It is a rapid grower, flourishing but a few feet from coal fire or radiator, in a north or south window. It quickly forgives neglect, and if allowed to dry up out of doors or indoors, recovers in due time when put in a moist atmosphere. It makes but one imperative demand, and that is the privilege of standing still. Overzealous culturists usually like to turn things around, but revolving cliffs are not in the natural order of things. The slender black stipes are very susceptible to changes of light and warped and twisted fronds result."

Dry, calcareous rocks, southern New England and westward. Rare. Var. cristata has forked pinnæ somewhat crowded toward the summit of the frond. Missouri.

(2) SMOOTH CLIFF BRAKE

Pellæa glabella. Pellæa atropurprea, var. Bushii

Naked with a few, scattered, spreading hairs, smooth surface and dark polished stipes. Rhizome short with membranous, orange or brown scales having a few bluntish teeth on each edge. Pinnæ sub-opposite, divergent, narrowly oblong, obtuse; base truncate, cordate or clasping, occasionally auricled; lower pinnæ often with orbicular or cordate pinnules. Sterile pinnæ broader, bluish or greenish glaucous above, often crowded to overlapping. The smooth cliff brake has a decidedly northern range, growing from northern Vermont to Missouri, and northwestward, but found rarely, if at all, in southern New England.

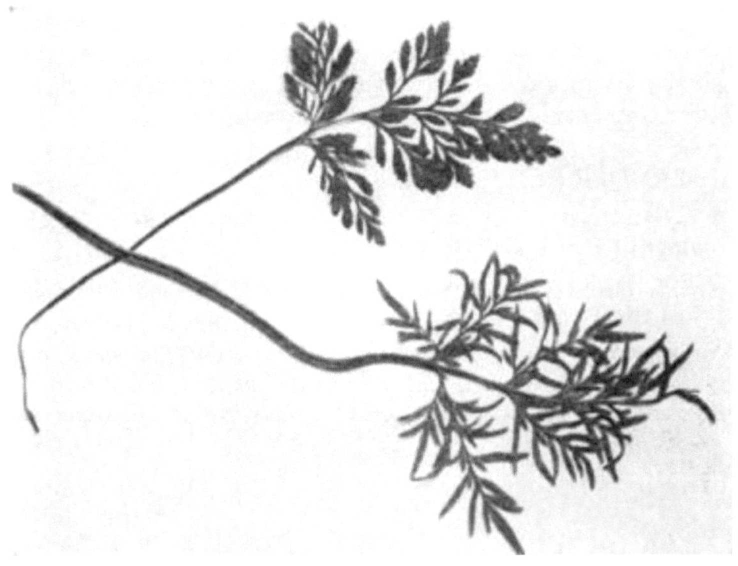

[Illustration: Dense Cliff Brake. Cryptogramma densa (From Waters's "Ferns," Henry Holt & Co.)]

(3) DENSE CLIFF BRAKE

Cryptogr ⋅ a densa. Pella衡densa

Modern botanists are inclined to place the dense cliff brake and the slender cliff brake under the genus Cryptogr ⋅ a, which is so nearly like Pellaea that one hesitates to choose between them. The word Cryptogr ⋅ a means in Greek a hidden line, alluding to the line of sporangia hidden beneath the reflexed margin.

The dense cliff brake may be described as follows:

Stipes three to nine inches tall, blades one to three inches, triangular-ovate, pinnate at the summit, and tripinnate below. Segments linear, sharp-pointed, mostly fertile, having the margins entire and recurved, giving the sori the appearance of half-open pods. Sterile fronds sharply serrate. Stipes in dense tufts ("densa") slender, wiry, light-brown.

This rare little fern is a northern species and springs from tiny crevices in rocks, preferring limestone. Like many other rock-loving species, it produces spores in abundance, having no other effective means of spreading, and its fertile fronds are much more numerous than the sterile ones, and begin to fruit when very small. Gasp頡nd Mt. Albert in the Province of Quebec, Grey County, Ontario, and in the far west.

(4) SLENDER CLIFF BRAKE

Cryptogr · a Stell版. Pella衡gr ·lis

Fronds (including stipes) three to six inches long, thin and slender with few pinn琪The lower pinn枆innately parted into three to five divisions, those of the fertile fronds oblong or linear-oblong; those of the sterile, obovate or ovate, crenulate, decurrent at the base. Confined to limestone rocks. Quebec and New Brunswick, to Vermont, Connecticut, Pennsylvania, and to the northwest.

[Illustration: Slender Cliff Brake. Cryptogramma Stelleri]

We have collected this dainty and attractive little fern on the limestone cliffs of Mt. Horr, near Willoughby Lake, Vt. It grew in a rocky grotto whose sides were kept moist by dripping water. How we liked to linger near its charming abode high on the cliff! And we liked also to speak of it by its pleasing, simple name, "Pell袴 gracilis," now changed for scientific reasons, but we still like the old name better.

(5) THE ROCK BRAKE. PARSLEY FERN

Cryptogr • a acrostich • s

Sterile and fertile fronds very dissimilar; segments of the fertile, linear and pod-like; of the sterile, ovate-oblong, obtuse, and toothed. The plants spring from crevices of rocks and are from six to eight inches high. Stipes of the fertile fronds are about twice as long as the sterile, making two tiers of fronds.

[Illustration: Parsley Fern or Rock Brake. Cryptogramma acrostichoides (California and Oregon) (Herbarium of Geo. E. Davenport)]

The parsley fern is the typical species of the genus Cryptogr • a. The in-dusium is formed of the altered margin of the pinnule, at first reflexed to *the midrib, giving it a pod-like appearance, but at length opening out flat and exposing the sporangia. Clute, speaking of this fern as "the rock brake," calls it a border species, as its home is in the far north--Arctic America to Lake Huron, Lake Superior, Colorado and California.*

4. LIP FERNS. Cheil▪hes

Mostly small southern ferns growing on rocks, pubescent or tomentose with much divided leaves. Sori at the end of the veins at first small and roundish, but afterwards more or less confluent. The indusium whitish and sometimes herbaceous, formed of the reflexed margin of the lobes or of the whole pinnule. Veins free, but often obscure. Most of the ferns of this genus grow in dry, exposed situations, where rain is sometimes absent for weeks and months. For this reason they protect themselves by a covering of hairs, scales or wool, which hinders the evaporation of water from the plant by holding a layer of more or less saturated air near the surface of the frond. (In Greek the word means lip flower, alluding to the lip-like indusia.)

(1) ALABAMA LIP FERN. Cheil▪hes alabam鳑is

Fronds smooth, two to ten inches long, lanceolate, bipinnate. Pinn桓umerous, oblong-lanceolate, the lower usually smaller than those above. Pinnules triangular-oblong, mostly acute, often auricular or lobed at the base. Indusia pale, membranous and continuous except between the lobes. Stipes black, slender and tomentose at the base.

[Illustration: Alabama Lip Fern. Cheilanthes alabamensis (From Waters's "Ferns," Henry Holt & Co.)]

This species of lip fern may be distinguished from all the others within our limits by its smooth pinn殒On rocks--mountains of Virginia to Kentucky, and Alabama, and westward to Arizona.

(2) HAIRY LIP FERN. Cheil · hes lan · C. v鳴ita

[Illustration: Hairy Lip Fern]

Fronds twice pinnate, lanceolate with oblong, pinnpfid pinnules; seven to fifteen inches tall, slender and rough with rusty, jointed hairs. Pinn捋 triangular-ovate, usually distant, the ends of the rounded lobes reflexed and forming separate involucres which are pushed back by the ripening sporangia.

This species like the other lip ferns is fond of rocks, springing from clefts and ledges. While hairy it is much less tomentose than the two following species. Unlike most of the rock-loving ferns this species is not partial to limestone, but grows on other rocks as well. It has been found as far north as New Haven, Conn., also near New York, and in New Jersey, Georgia, and westward to Wyoming and southward.

(3) WOOLLY LIP FERN. Cheil • hes toment •/i>

Fronds eight to eighteen inches long, lanceolate-oblong, tripinnate. Pinn校nd pinnules ovate-oblong, densely woolly especially beneath, with slender, whitish, obscurely jointed hairs. Of the ultimate segments the terminal one is twice as long as the others. Pinnules distant, the reflexed, narrow margin forming a continuous, membranous indusium. Stipe stout, dark brown, densely woolly.

By donning its thick coat of wool this species is prepared to grow in the most exposed situations of the arid southwest. It is said to be the "rarest, tallest and handsomest of the lip ferns."

Mountains of Virginia and Kentucky to Georgia, and west to Missouri, Texas and Arizona.

(4) SLENDER LIP FERN

Cheil • hes F饯 C. lanugin •/i>

Stipes densely tufted, slender, at first hairy, dark brown, shining. Fronds three to eight inches long, ovate-lanceolate, with thickish, distinctly articulated hairs, twice or thrice pinnate. Pinn柘vate, the lowest deltoid. Pinnules divided into minute, densely crowded segments, the herbaceous margin recurved and forming an almost continuous indusium.

[Illustration: Slender Lip Fern]

The slender lip fern, known also as Fée's fern, is much the smallest of the lip ferns, averaging, Clute tells us, "but two inches high." This is only one-third as tall as the woolly lip fern and need not be mistaken for it. The fronds form tangled mats difficult to unravel. It grows on dry rocks and cliffs--Illinois and Minnesota to British Columbia, and south to Texas, New Mexico and Arizona.

[Illustration: Pinn柄f Slender Lip Fern. Cheilanthes F饩 (From Waters's "Ferns," Henry Holt & Co.)]

5. CLOAK FERN. Notholⵎa

Small ferns with fruit-dots borne beneath the revolute margin of the pinnules, at first roundish, but soon confluent into a narrow band without indusium. Veins free. Fronds one to several times pinnate, the lower surface hairy, or tomentose or powdery. Includes about forty species, mostly American, but only one within our limits. (Greek name means spurious cloak, alluding to the rudimentary or counterfeit indusium.)

(1) POWDERY CLOAK FERN. *Notholæa dealbæ*

Fronds two to six inches long, triangular-ovate, acute, broadest at the base, tripinnate. Stalks tufted, wiry, shining, dark brown. Upper surface of the very small segments green, smooth, the lower densely coated with a pure, white powder; hence, the specific name dealbata, which means whitened. Sori brown at length; veins free.

There are several species of cloak ferns, but only one within our limits. The dry, white powder which covers them doubtless is designed to protect them from too rapid evaporation of moisture, as they all inhabit dry and sunny places. This delicate rock-loving fern is found in the clefts of dry limestone rocks in Missouri, Kansas, Colorado, and southwestward.

[Illustration: Powdery Cloak Fern. Notholaena dealbata (Kansas) (G.E. Davenport)]

THE CHAIN FERNS. *Woodwardia*

Large and somewhat coarse ferns of swampy woods with pinnate or nearly two-pinnate fronds, and oblong or linear fruit-dots, arranged in one or more chain-like rows, parallel to and near the midribs. Indusium fixed by its outer margin to a veinlet and opening on the inner side. In our section there are two species. (Named for Thomas J. Woodward, an English botanist.)

(1) THE COMMON CHAIN FERN. *Woodwardia virgica*

Sterile and fertile fronds similar in outline, two to four feet high, once pinnate, the pinnæ deeply incised with oblong segments. Fruit-dots oblong in chain-like rows along the midrib both of the pinnæ and the lobes, confluent when ripe. Veins forming narrow rows of net-like spaces (areoles) beneath the fruit-dots, thence free to the margin. The spores ripen in July.

[Illustration: The Common Chain Fern. Woodwardia virginica]

The sterile fronds resemble those of the cinnamon fern, but the latter grow in crowns, with a single frond in the center, while the fronds of the chain fern rise singly from the creeping rootstock, which sends them up at intervals all summer. The sori are borne on the backs of fertile fronds. There are usually more sterile than fertile blades, especially in dense shade. We have waded repeatedly through a miry swamp in Melrose, Mass., where the wild calla flourishes along with the blueberry and other swamp bushes, and have found the chain fern in several shaded spots, but every frond was sterile. It is said that when exposed to the sun it always faces the south. Swamps, Maine to Florida, especially along the Atlantic Coast, and often in company with the narrow-leaved species.

[Illustration: Net-Veined Chain Fern. Woodwardia areolata (Stratford, Conn.)]

(2) NET-VEINED CHAIN FERN

NARROW-LEAVED CHAIN FERN

Woodwardia areolƜ. W. angustif□

Root stocks creeping and chaffy. Sterile and fertile fronds unlike; sterile ones nine to twelve inches tall, deltoid-ovate. Broadest at the base, with lanceolate, serrulate divisions united by a broad wing. Veins areolate; fertile fronds taller, twelve to twenty inches high with narrowly linear divisions, the areoles and fruit-dots in a single row each side of the secondary midrib, the latter sunk in the tissues.

This species is less common than the Virginia fern, but they often grow near each other. We have collected both in the Blue Hill reservation near Boston, and both have been found in Hingham, Medford, and Reading, and doubtless in other towns along the coast. Mrs. Parsons speaks of finding them in the flat, sandy country near Buzzard's Bay. The net-veined species has some resemblance to the sensitive fern, but in the latter the spore cases are shut up in small pods formed by the contracting and rolling up of the lobes, whereas the chain fern bears its sori on the under side of long, narrow pinn殯Besides, the sterile fronds of the latter have serrulate segments. As in the sensitive fern there are many curious gradations between the fertile and sterile fronds, both in shape and fruitfulness. Waters calls them the "obtusilobƜ form."

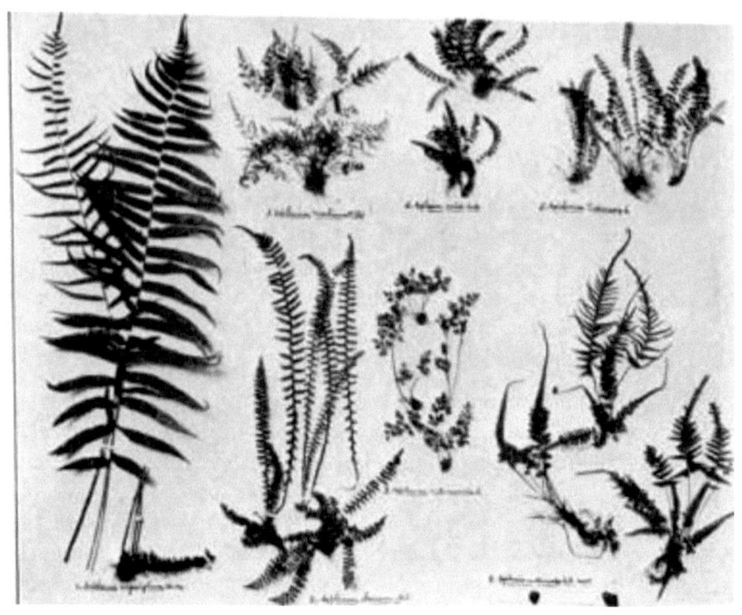

[Illustration: The Spleenworts 1. Narrow-leaved 2. Ebony 3. Rue 4. Scott's 5. Maidenhair 6. Green 7. Mountain]

THE SPLEENWORTS

*A. THE ROCK SPLEENWORTS. **Aspl ìĿum***

Small, evergreen ferns. Fruit-dots oblong or linear, oblique, separate when young. Indusium straight or rarely curved, fixed lengthwise on the upper side of a fertile veinlet, opening toward the midrib. Veins free. Scales of rhizome and stipes narrow, of firm texture and with thick-walled cells.

*(1) PINN¶IFID SPLEENWORT. **Asplenium pinnat • dum***

Fronds four to six inches long, lanceolate, pinnᴘfid or pinnate near the base, tapering above into a slender prolongation. Lobes roundish-ovate, or the lower pair acuminate. Fruit-dots irregular, numerous. Stipes tufted, two to four inches long, brownish beneath, green above.

Although this fern, like all the small spleenworts, is heavily fruited, it is extremely rare. It is found as far north as Sharon, Conn., thence southward to Georgia, to Arkansas and Missouri. On cliffs and rocks. Resembles the walking fern, and its tip sometimes takes root.

*(2) SCOTT'S SPLEENWORT. **Asplenium eben • s***

Fronds four to ten inches long, broadly lanceolate, pinnᴘfid or pinnate below, tapering to a prolonged and slender apex. Divisions lanceolate from a broad base. Fruit-dots straight or slightly curved. Stipe and rachis brown.

[Illustration: Pinnatifid Spleenwort. Asplenium pinnatifidum a, Small Plants from Harper's Ferry; b, Sori on Young Fronds (From Waters's "Ferns," Henry Holt & Co.)]

[Illustration: Scott's Spleenwort. Asplenium ebenoides a, from Virginia; b, from Alabama; c, from Maryland (From Waters's "Ferns," Henry Holt & Co.)]

Resembles the last, and like that has been known to root at the tip. It is a hybrid between the walking fern and the ebony spleenwort, as proved by Miss Margaret Slosson, and may be looked for in the immediate vicinity of its parents. It was discovered by R.R. Scott, in 1862, at Manayunk, Pa., a suburb of Philadelphia, and described by him in the Gardener's Monthly of September, 1865. Vermont to Alabama, Missouri, and southward. Rare, but said to be plentiful in a deep ravine near Havana, Ala.

[Illustration: Green Spleenwort. Asplenium v□de]

(3) GREEN SPLEENWORT. *Asplenium v□de*

Fronds two to ten inches long, linear, pinnate, pale green. Pinn□roundish-ovate, crenate, with indistinct and forking midveins. Stalks tufted, short, brownish below, green above. Rachis green.

Discovered at Smuggler's Notch, Mt. Mansfield, Vt., by C.G. Pringle in 1876. Found sparingly at Willoughby Lake, high on the cliffs of Mt. Horr. This rare and delicate little plant bears a rather close resemblance to the maidenhair spleenwort, which, however, has dark stipes instead of green.

Northern New England, west and northwest on shaded limestone rocks.

[Illustration: Maidenhair Spleenwort. Asplenium Trichomanes]

(4) MAIDENHAIR SPLEENWORT. *Asplenium Trich□es*

Stipes densely tufted, purple-brown, shining. Fronds three to eight inches long, linear, dark green, rather rigid. Pinn楮oundish-oblong or oval, entire or finely crenate, attached at the base by a narrow point. Midveins forking and evanescent.

Not very common, but distributed almost throughout North America. May be looked for wherever there are ledges, as it does not require limestone. July.

[Illustration: Maidenhair Spleenwort. Asplenium Trichomanes (From Woolson's "Ferns," Doubleday, Page & Co.)]

(5) SMALL SPLEENWORT

Asplenium p□ulum. A. res□ens

Fronds four to ten inches tall, narrowly linear, rather firm, erect. Pinn□ opposite, oblong, entire or finely crenate, and auricled at the base. Stipes and rachis black and shining. Midveins continuous.

This small fern is a southern species half way between the maidenhair and ebony spleenworts, but rather more like the latter from which it differs in being smaller and thicker, and in having the fertile and sterile fronds of the same size. Mountains of Virginia to Kansas and southward.

(6) EBONY SPLEENWORT

Asplenium platyn□on. A. eb□um

Fronds upright, eight to eighteen inches high, linear-lanceolate, the fertile ones much taller, and pinnate. Pinn□carcely an inch long, the lower ones very much shorter, alternate, spreading, finely serrate or incised, the base auricled. Sori numerous, rather near the midvein, stipe and rachis lustrous brown. ("Ebony.")

94

This rigidly upright but graceful fern flourishes in rocky, open woods, and on rich, moist banks, often in the neighborhood of red cedars. Having come upon it many times in our rambles, we should say it was not un-common.

A lightly incised form of the pinnæ has been described as var. serratum. A handsome form discovered in Vermont in 1900 by Mrs. Horton and named Hortoni> (also called incisum) has plume-like fronds with the pinnæ cut into oblique lobes, which are coarsely serrate.

[Illustration: Ebony Spleenwort. Asplenium platyneuron (Melrose, Mass., G.E. Davenport)]

[Illustration: Bradley's Spleenwort. Asplenium Bradleyi a, from Maryland; b, from Kentucky (From Waters's "Ferns," Henry Holt & Co.)]

(7) BRADLEY'S SPLEENWORT. *Asplenium Br□eyi*

Fronds oblong-lanceolate, pinnate, three to ten inches long. Pinn楠blong-ovate, obtuse, incised or pinnₚfid into oblong, toothed lobes. The basal pinn枨ave broad bases, and blunt tips and are slightly stalked. Stipes and rachis dark brown and the sori short, near the midrib.

A rare and beautiful fern growing on rocks preferring limestone and confined mostly to the southern states. Newburg, N.Y., to Kentucky and Alabama, westward to Arkansas.

(8) MOUNTAIN SPLEENWORT. *Asplenium montanum*

Fronds ovate-lanceolate from a broad base, two to eight inches long, somewhat leathery, pinnate. Pinn柄vate-oblong, the lowest pinnately cleft into oblong or ovate cut-toothed lobes, the upper ones less and less divided. Rachis green, broad, and flat.

[Illustration: Mountain Spleenwort (From the "Fern Bulletin")]

Small evergreen ferns of a bluish-green color, growing in the crevices of rocks and cliffs. Connecticut to Ohio, Kentucky, Arkansas and southwest. July. Rare. Williams, in his "Ferns of Kentucky," says of this species, "Common on all sandstone cliffs and specimens are large on sheltered rocks by the banks of streams."

(9) RUE SPLEENWORT. *Asplenium Ruta-mur*□*a*

Fronds evergreen, small, two to seven inches long, deltoid-ovate, two to three pinnate below, simply pinnate above, rather leathery in texture. Divisions few, stalked, from cuneate to roundish-ovate, toothed or incised at the apex. Veins forking. Rachis and stipe green. Sori few, soon confluent.

[Illustration: The Rue Spleenwort. A. Ruta-muraria (Top, Lake Huron--Lower Left, Mt. Toby, Mass.--Lower Right, Vermont) (From Herbarium of Geo. E. Davenport)]

This tiny fern grows from small fissures in the limestone cliffs, and is rather rare in this country; but in Great Britain it is very common, growing everywhere on walls and ruins. From Mt. Toby, Mass., and Willoughby Mountain, Vt., to Michigan, Missouri, Kentucky and southward.

B. THE LARGE SPLEENWORTS. *Athrium*

The following species, which are often two to three feet high and grow in rich soil, are quite different in appearance and habits from the small rock spleenworts just described. Some botanists have kept them in the genus Asplenium because their sori are usually rather straight or only slightly curved, but others are inclined to follow the practice of the British botanists and put them into a separate group under Athrium. Nearly all agree that the lady fern, with its variously curved sori, should be placed here, and many others would place the silvery spleenwort in the same genus, partly because of its frequently doubled sori. In regard to the last member of the group, the narrow-leaved spleenwort, there is more doubt. The sori taken separately would place it with the Aspleniums, but considering its size, structure, habits of growth and all, it seems more closely allied to the two larger ferns than to the little rock species. We shall group the three together as the large spleenworts, or for the sake of being more definite adopt Clute's felicitous phrase.

THE LADY FERN AND ITS KIN

1. THE LADY FERNS

Fronds one to three feet high, broadly lanceolate, or ovate-oblong, tapering towards the apex, bipinnate. Pinnæ lanceolate, numerous. Pinnules oblong-lanceolate, cut-toothed or incised. Fruit-dots short, variously curved. Indusium delicate, often reniform, or shaped like a horseshoe, in some forms confluent at maturity.

Widely distributed, common and varying greatly in outline. The newer nomenclature separates the lady fern of our section into two distinct species, which should be carefully studied.[A]

[Footnote A: See monograph by F.K. Butters in Rhodora of September, 1917.]

(1) THE UPLAND LADY FERN. ATHYRIUM ANGUSTUM

Asplenium Filix-femina

The rootstock or rhizome of the Upland Lady Fern here pictured shows how the thick, fleshy bases of the old fronds conceal the rootstock itself. In the Lowland Lady Fern the rootstock is but slightly concealed by old stipe bases, and so may be distinguished from its sister fern.

One design of such rootstocks is to store up food (mostly starch), during the summer to nourish the young plants as they shoot forth the next spring. The undecayed bases of the old stipes are also packed with starch for the same purpose.

[Illustration: Rootstock of the Upland Lady Fern]

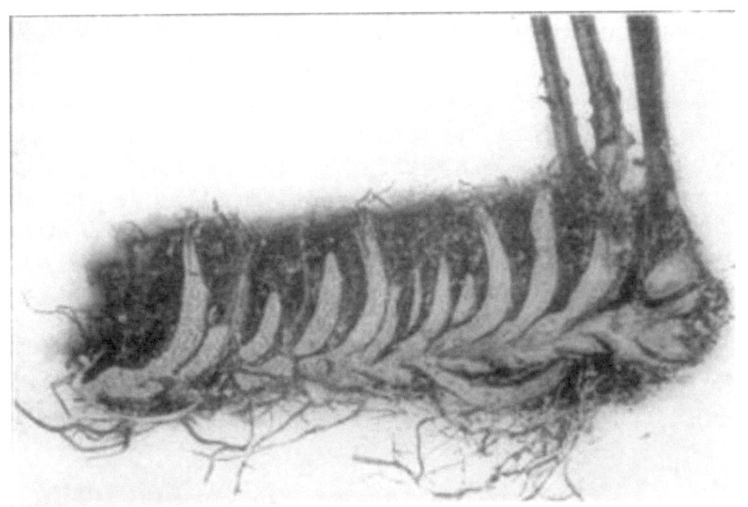

[Illustration: The same split lengthwise (From Waters's "Ferns," Henry Holt & Co.)]

[Illustration: Sori of Lady Fern. Athyrium angustum]

Rootstocks horizontal, quite concealed by the thick, fleshy bases of old fronds. Scales of the long, tufted stipes dark brown. Indusium curved, often horseshoe-shaped, usually toothed or fringed with fine hairs, but without glands. Fronds bipinnate, one to three feet high, widest near the middle.

This is the common species of northern New England and the Canadian Provinces. The fronds differ very widely in form and a great many varieties have been pointed out, but the fern student, having first learned to identify the species, will gradually master the few leading varieties as he meets them.

Those growing in warm, sunny places where the fruit-dots when mature incline to cover the whole back of the frond are called "sun forms." These are varieties TYPICUM and ELATIUS, both with the pinnæ obliquely ascending (including variety angustum of D.C. Eaton), but the latter has broader fronds with the pinnules of the sterile fronds oblong-lanceolate, somewhat acute and strongly toothed or pinnatifid.

[Illustration: Varieties of Lady Fern Left to right--lst and 2nd, Var. typicum; 3d, elatius; 4th, rubellum; 5th, uncertain, perhaps confertum]

Var. RUBRLUM has the sori distinct even when mature; its pinnules stand at a wide angle from the rachis of the pinna and are strongly toothed or pinnatifid with obtuse teeth. This variety favors regions with cool sum-

mers, or dense shade in warmer regions. The term RUBRLUM alludes to the reddish stems so often seen but this sign alone may not determine the variety. It occurs throughout the range of the species, being a common New England fern. Fernald remarks that this is also a common form of the species in southern Nova Scotia.

Among other varieties named by Butters are CONFvTUM, having the pinnules irregularly lobed and toothed; joined by a membranous wing, the lobes of the pinnules broad and overlapping, giving the fern a compact appearance; LACINI⁑UM with pinnules very irregular in size and shape, with many long, acute teeth, which project in various directions. "An abnormal form which looks as if it had been nibbled when young."

These varieties are represented in the Gray Herbarium.

(2) THE LOWLAND LADY FERN

ATHꞋIUM ASPLENIꞋDES

Rootstocks creeping, not densely covered with the persistent bases of the fronds. Stipes about as long as the blade. Scales of the stipe very few, seldom persistent, rarely over 3-16 of an inch long. Fronds narrowly deltoid, lanceolate, widest near the base, the second pair of pinn⫽ommonly longest. Indusia ciliate, the cilia (hairs) ending in glands. Spores dark, netted or wrinkled.

[Illustration: Lowland Lady Fern. ATHYRIUM ASPLENIOIDES (From the Gray Herbarium)]

The following two forms are named by Butters:

F. TYPICUM. The usual form frequent in eastern Massachusetts, Rhode Island, Connecticut, New York, Pennsylvania, North Carolina, Ohio, and Missouri.

F. SUBTRIPINNATUM. An unusually large and rare form with triangular, lanceolate, and pinnatifid pinnules, having blunt, oblong segments. Wet situations in half shade. Massachusetts, West Virginia, and Virginia.

Our lowland or southern lady fern flourishes in the southern states, comes up the Atlantic Coast until it meets the upland or northern species in Pennsylvania and southern New England, and their identification can hardly fail to awaken in the student a keen interest.

Our American botanists are inclined to think that the real Athrium filix-femina is not to be found in the northeastern United States, but is rather a western species, with its habitat in California and the Rocky Mountain region and identical with Athrium cyclos ·.

But whatever changes may occur in the scientific name of the old Athrium filix-femina, the name lady fern will not change, but everywhere within our limits it will hold its own as a familiar term.

Underwood, writing of the lady fern under the genus Asplenium, mentions the form "exile, small, starved specimens growing in very dry situations and often fruiting when only a few inches high." He also mentions Eaton's "angstum," and alludes to the "Remaining sixty-three varieties equally unimportant that have been described of this species."

The lady fern is common in moist woods, by walls and roadsides, and at its best is a truly handsome species, although, like Mrs. Parsons, we have noticed that in the late summer it loses much of its delicacy. "Many of its forms become disfigured and present a rather blotched and coarse appearance." The lady fern has inspired several poems, which have been quoted more or less fully in the fern books. The following lines are from the pen of Calder Campbell:

"But not by burne in wood or dale Grows anything so fair As the palmy crest of emerald pale Of the lady fern when the sunbeams turn To gold her delicate hair."

Referring, perhaps, to the fair colors of the unfolding crosiers revealing stipes of a clear wine color in striking contrast with the delicate green of the foliage.

In identifying this fern the novice should bear in mind the tendency of the curved sori of youth to become straightened and even confluent with age, although such changes are rather unreliable. Possibly the suggestion of the poetic Davenport may be helpful to some that there is "An indefinable charm about the various forms of the lady fern, which soon enables one to know it from its peculiarly graceful motion by merely gently swaying a frond in the hand." Spores ripen in August.

The lady fern is very easy to cultivate and when once established is apt to crowd aside its neighbors.

(3) SILVERY SPLEENWORT. ATHYRIUM ACROSTICHOIDES

Asplenium acrostich•s. Asplenium thelypter•s

Fronds two to four feet tall, pinnate, tapering both ways from the middle. Pinnæ deeply pinnatifid, linear-lanceolate, acuminate. Lobes oblong, obtuse, minutely toothed, each bearing two rows of oblong or linear fruit-dots. Indusium silvery when young.

[Illustration: Silvery Spleenwort. Athyrium acrostichoides]

[Illustration: Silvery Spleenwort. Athyrium acrostichoides]

The sterile fronds come up first and the taller, fertile ones do not appear until late in June. Where there are no fruit-dots the hairs on the upper surface of the fronds will help to distinguish it from specimens of the Marsh fern tribe, which it somewhat resembles. The regular rows of nearly straight, clear-cut sori of the fertile fronds are very attractive, and the lower ones, as well as those at the slender tips of the pinnæ are frequently double.

Rich woods and moist, shady banks, New England to Kentucky and westward. Generally distributed but hardly common.

(4) NARROW-LEAVED SPLEENWORT

ATHYRIUM ANGUSTIFOLIUM. Asplenium angustifolium

Fronds one to four feet tall, pinnate. Pinnæ numerous, thin, short-stalked, linear-lanceolate, acuminate, those of the fertile fronds narrower. Fruit-dots linear. Indusium slightly convex.

[Illustration: Narrow-leaved Spleenwort. Athyrium angustifolium (Vermont) (Geo. E. Davenport)]

In rich woods from southern Canada and New Hampshire to Minnesota and southward. September. Not common. Mt. Toby, Mass., Berlin and Meriden, Conn., and Danville, Vt. Can be cultivated but should not be exposed to severe weather, as its thin and delicate fronds are easily injured. Woolson writes of it, "There is nothing in the fern kingdom which looks so cool and refreshing on a hot day as a mass of this clear-cut, delicately made-up fern."

[Illustration: Pinn校nd Sori of Athyrium angustifolium]

HART'S TONGUE

Scolop裂rium. PHYLLIS

Sori linear, a row on either side of the midvein, and at right angles to it, the indusium appearing to be double. (Scolopendrium is the Greek for centipede, whose feet the sori were thought to resemble. Phyllitis is the ancient Greek name for a fern.) Only one species in the United States.

(1) *Scolopendrium vulgœ*

PHYLLIS SCOLOPYDRIUM

Fronds thick and leathery, oblong-lanceolate from an auricled, heart-shaped base, ten to twenty inches long and one to two inches wide. Margin entire, bright green.

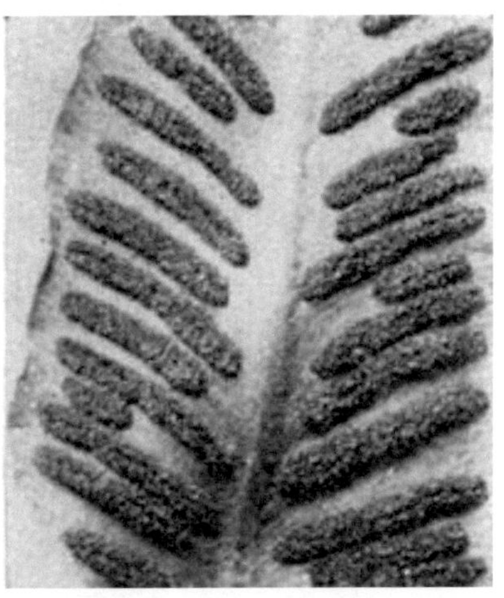

[Illustration: Sori of *Scolopendrium vulgare*]

In shaded ravines under limestone cliffs. Chittenango Falls, and Scolopendrium Lake, central New York, and Tennessee. Also, locally in Ontario and New Brunswick. One of the rarest of our native ferns, although very common in Great Britain. This plant is said to be easily cultivated, and to produce numerous varieties. According to Woolson, "No rockery is complete without the Hart's Tongue, the long, glossy, undulating fronds of which are sufficiently unique to distinguish any collection." In cultivation it "needs light protection through the winter in northern New England."

[Illustration: Hart's Tongue. *Scolopendrium vulgare* (Base of calcareous rocks, Owen Sound, Ontario, Canada)]

WALKING FERN. WALKING LEAF

Camptos☐

Fruit-dots oblong or linear as in Aspl i um, but irregularly scattered on either side of the reticulated veins of the simple frond, the outer ones sometimes confluent at their ends, forming crooked lines (hence, the name from the Greek meaning crooked sori). Only one species within our limits.

Camptos • rhizophyllus

Fronds evergreen, leathery, four to eighteen inches long, heart-shaped at the base, but tapering towards the apex, which often roots and forms a new plant. Veins reticulated. The auricles of this species are sometimes elongated and may even take root.

This curious and interesting fern is one of the finest for rockeries, the tips taking root in rock-fissures. Shaded limestone, or sometimes other rocks. Shapleigh and Winthrop, Me., rarely in New Hampshire (Lebanon), and Connecticut, Mt. Toby, Mass., and western New England; also Canada to Georgia and westward.

[Illustration: Walking Fern. *Camptosorus rhizophyllus*]

THE SHIELD FERNS

THE CHRISTMAS AND HOLLY FERNS

Polstichum

These have been grouped with the wood ferns, but are now usually placed under the genus Polstichum, which has the sori round and covered with a circular indusium fixed to the frond by its depressed center. The wood ferns, on the other hand, have a kidney-shaped indusium attached to the fronds by the sinus. (Polstichum is the Greek for many rows, the sori of some species being in many ranks.)

(1) THE CHRISTMAS FERN

Polstichum acrostichœs. Aspⁱum acrostichœs

Stipes clothed with pale, brown scales. Frond rigid and evergreen, one to two feet long, lanceolate, pinnate. Pinnæ linear-lanceolate, scythe-shaped, auricled on the upper side, and with bristly teeth; fertile pinnæ contracted toward the top, bearing two rows of sori, which soon become confluent and cover the entire surface. Indusium orbicular, fixed by its depressed center.

F. incisum is a form in which the pinnæ are much incised.

F. cr•um has the edges of its pinnæ crisped and ruffled. The name Christmas fern, due to John Robinson, of Salem, Mass., suggests its fitness for winter decoration. Its deep green and glossy fronds insure it a welcome at Christmas time. "Its mission is to cheer the winter months and enhance the beauty of the other ferns by contrast." In transplanting, a generous mass of earth should be included and its roots should not be disturbed.

[Illustration: Christmas Fern. Polystichum acrostichoides]

[Illustration: Christmas Fern. Polystichum acrostichoides]

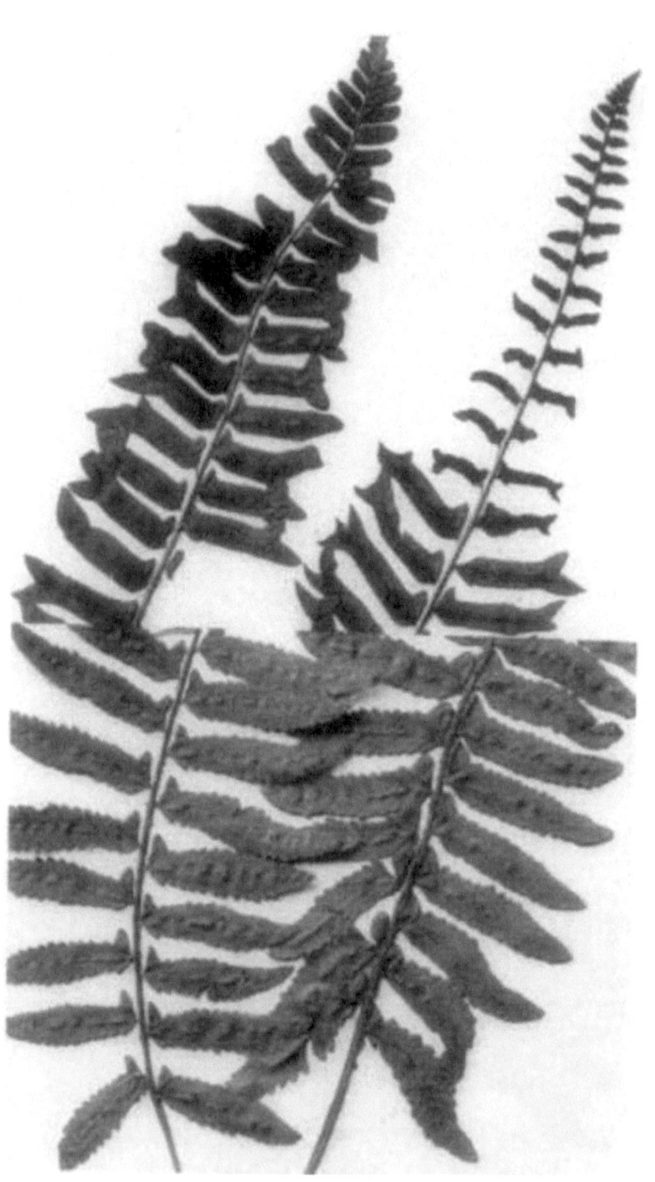

[Illustration: Christmas Fern. Polystichum acrostichoides Top, Forked Form; Bottom, Incised Form (Maine)]

(2) BRAUN'S HOLLY FERN

Polystichum Brſii. Asp☐um acule☾ſn Brſii

Fronds thick, rigid, one to two feet long, spreading, lanceolate, tapering both ways, bipinnate. Pinnules ovate or oblong, truncate, nearly rectangular at the base, sharply toothed and covered beneath with chaff and hairs. Fruit-dots small and near the mid veins. Indusium orbicular, entire. Stipes chaffy with brown scales.

[Illustration: Braun's Holly Fern. Polystichum Braunii (Willoughby Mountain, Vt.) (Herbarium of G.H.T.)]

This handsome fern is rather common in northern New England. We have collected it in the Willoughby Lake region, Vt., and it is found at Mt. Mansfield, Randolph, and elsewhere in that state; also at Gorham, N.H., and Fernald reports it as common in northern Maine. It also grows in the mountains of New York and Pennsylvania, and westward. It was formerly thought to be a variety of the prickly shield fern (P. acule☉m), which has a very wide range and numerous varieties. The fronds remain green through the winter but the stipes weaken and fall over.

(3) HOLLY FERN. *Polystichum Lonch☒s*

Fronds linear-lanceolate, short-stalked and rigid, eight to fifteen inches long. Pinn☒roadly lanceolate-falcate or the lowest triangular, strongly auricled on the upper side, densely spinulose-toothed. Sori midway between the margin and midrib.

The name holly fern suggests its resemblance to holly leaves with their bristle-tipped teeth. The specific name lonchēs (like a spear) refers to its sharp teeth. A northern species growing in rocky woods from Labrador to Alaska, and south to Niagara Falls, Lake Superior and westward. Its southern limits nearly coincide with the northern limits of the Christmas fern.

THE MARSH FERN TRIBE

Under this designation Clute has grouped three of the shield ferns, which have a close family resemblance, and has thus distinguished them from the wood ferns, which also belong to the shield fern family.

(1) THE MARSH FERN

Asp☐um thelpteris. THEL ⌣TERIS PAL⸲TRIS

Dry • ris thelpteris. Nephr • m thelpteris

[Illustration: The Marsh Fern]

These are all good names and each one is worthy to be chosen. Asp • um, Greek for shield, in use for a century, adopted in all the seven editions of Gray's Manual, is still the most familiar and pleasing term to its friends. Dry • ris, Greek for oak fern, has been chosen by Underwood and Britton and Brown and has grown in favor. Nephr • m, meaning kidney-like, favored by Davenport, Waters and, of late, Clute, is a most fitting name. THEL⌣TERIS, meaning lady fern, is found to be the earliest name in use and according to rule the correct one.

[Illustration: The Marsh Fern. Aspidium Thelypteris]

Fronds pinnate, lanceolate, slightly or not at all narrowed at the base. Pinnæ horizontal or slightly recurved, linear-lanceolate and deeply pinnatifid. Lobes obtuse, but appear acute when their margins are reflexed over the sori. Veins once forked. Indusium minute. Stipes tall, lifting the blades ten to fifteen inches above the mud, whence they spring.

The fronds of the marsh fern are apt to be sterile in deep shade. It may be readily distinguished from the New York fern by its broad base, instead of tapering to very small pinnæ, by its long stalk, lifting the blade up into the sunlight, and by the revolute margins of the fertile fronds, which have suggested for it the name of "snuff-box" fern. It is separated from the Massachusetts fern by its forked veins. Common in marshes and damp woodlands; Canada to Florida and westward. While the marsh fern loves moisture and shade it is sometimes found in dry, open fields. Miss Lilian A. Cole, of Union, Me., reports a colony as growing on land above the swale in which Twayblade and Adder's Tongue are found, "around rock heaps in open sunlight on clay soil, but homely and twisted," as if a former woodsy environment had been long since cleared away while the deserted ferns persisted.

(2) MASSACHUSETTS FERN

Aspidium simulatum. THELYPTERIS SIMULATA

Dryopteris simulata. Nephrodium simulatum

Fronds pinnate, one to three feet long, oblong-lanceolate, somewhat narrowed at the base. Pinnæ lanceolate, deeply pinnatifid, the lower most often turned inward. Veins simple. Indusium glandular. Sori rather large.

Resembles the marsh fern, of which it was once thought to be a variety. In some respects it is also like the New York fern, and is in fact intermediate between the two.

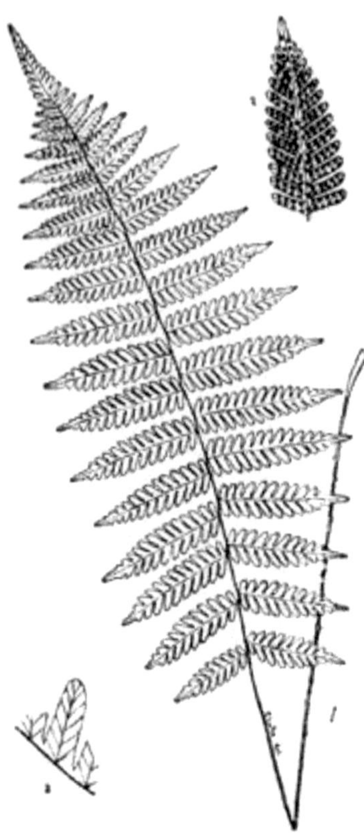

[Illustration: Massachusetts Fern. *Aspidium simulatum* 1. Sterile Frond. 2. A Fruiting Pinnule. 3. Pinnule enlarged showing venation (From the "Fern Bulletin")]

That it is a distinct species was first pointed out by Raynal Dodge in 1880, and it later was named simul⌒Jm> by Geo. E. Davenport because of its similarity to a form of the lady fern. It may be identified by its thin texture and particularly by its simple veins. On account of its close resemblance to the marsh fern, Clute would call it "The lance-leaved Marsh Fern," instead of the irrelevant name of Massachusetts Fern. Woodland swamps usually in deep shade, New England to Maryland and westward. Often found growing with the marsh fern.

(3) NEW YORK FERN

Aspidium noveborac鲔e. THEL⌣TERIS NOVEBORACΥSIS

Dry·ris noveborac鲔is. Nephr·m noveborac鲔e

Fronds pinnate, tapering both ways from the middle. Pinn槌anceolate, pinnatifid, the lowest pairs gradually shorter and deflexed. Veins simple. Indusium minute and beset with glands.

[Illustration: New York Fern. Aspidium noveboracense]

Very common in woodlands, preferring a dryer soil than the marsh fern. August. The fronds are pale green, delicate and hairy beneath along the midrib and veins.

[Illustration: Sori of New York Fern (From Waters's "Ferns," Henry Holt & Co.)]

[Illustration: New York Fern. Aspidium noveboracense]

When bruised its resinous glands give out a pleasing, ferny odor. This species can be distinguished from every other by the greatly reduced pinn校t its base. Throughout North America east of the Mississippi.

THE BEECH FERNS

*The beech ferns are often classed with the polypodies, because, like them, they have no indusium; but in other ways they are more akin to the wood ferns. Their stipes are not jointed to the root stock, nor are their sori at the ends of the veins as in the polypodies. We here place them with the wood ferns, retaining the familiar name Pheg*ris *but giving THEL*TERIS *as a synonym. The fruit-dots are small, round and naked, borne on the back of the veins below the apex. Stipe continuous with the rootstock. Veins free. (The name Pheg · ris in Greek means oak or beech fern.)*

(1) OAK FERN

Pheg · ris dry · ris. THELTERIS DRYĂTERIS

Fronds glabrous, broadly triangular, ternate, four to seven inches broad, the divisions widely spreading, each division pinnate at the base. Segments oblong, obtuse, entire or toothed. Fruit-dots near the margin. Rootstock slender and creeping from which fronds are produced all summer, in appearance like the small, ternate divisions of the bracken.

This dainty fern has fronds of a delicate yellow-green, "the greenest of all green things growing." Its ternate character is shown even in the uncoiling of the fronds, the three round balls suggesting the sign of the pawnbroker. The parts of the oak fern develop with great regularity, each pinna, pinnule and lobe having another exactly opposite to it nearly always. In rocky woods, common northward; also in Virginia, Kansas and Colorado. A fine species for cultivation at the base of the artificial rockery.

[Illustration: Oak Fern. Phegopteris Dryopteris]

(2) THE NORTHERN OAK FERN

Phegopteris Robertiana. Phegopteris calcarea

THELYPTERIS ROBERTI

Resembles the oak fern, but with fronds rather larger, especially the terminal segment; also more rigid and coarser in appearance. Stalks and fronds minutely glandular beneath. Lower pinnules of the lateral divisions scarcely longer than the others. Often called "Limestone Polypody," the beech ferns having formerly been classed with the polypodies. Britton and Brown designate it as the "Scented Oak Fern." Canada and the northwestern states. Rare.

[Illustration: Northern Oak Fern. Phegopteris Robertiana (From Water's "Ferns," Henry Holt & Co.)]

[Illustration: BROAD BEECH FERN. *Phegopteris hexagonoptera*]

(3) BROAD BEECH FERN

Phegopteris hexagon☐ra

THEL◡TERIS HEXAGONĂTERA

Fronds triangular, broader than long, seven to twelve inches broad, spreading more or less horizontally at the summit of the stipe; pubescent and often glandular beneath; pinn☒ragrant, lanceolate, the lowest pair usually much larger than those above, having the segments elongated and cut into lobes. Basal segments decurrent and forming a many-angled wing along the main rachis. Fruit-dots small, near the margin.

The broad beech fern is usually larger than its sister, the long beech fern, and extends farther south, ranging from New England to Minnesota and southward to Florida. It is sometimes called "six-angled polypody." According to Dodge it is most common in Rhode Island and Connecticut. It prefers rather dry, open woods. It is said to have a pleasant, ferny odor when bruised. August.

(4) LONG BEECH FERN

Phegopteris polypodi •s. THEL◡TERIS PHEGĂTERIS

Fronds triangular, longer than broad, four to six inches long, twice pinnatifid. Pinn☒anceolate, acuminate, the lowest pair deflexed and standing forward; cut into oblong, obtuse segments. Fruit-dots near the margin.

Compared with the broad beech fern this is the more northern species. While usually quite distinct in structure, it sometimes approaches its sister fern rather closely.

It prefers deep woods and shaded banks. Newfoundland to Alaska and southward to the mountains of Virginia. July.

[Illustration: Long Beech Fern. Phegopteris polypodioides]

[Illustration: The Long Beech Fern]

THE FRAGRANT FERN

Asp☐um fr ꞈans. Nephr☐m fr ꞈans

THEL◡TERIS FR●RANS. Dry · ris fr ꞈans

Fronds four to twelve inches high, glandular-aromatic, narrowly lanceolate and twice pinnate or nearly so. Pinn☐oblong-lanceolate, pinnate or deeply pinnatifid. Pinnules toothed or entire nearly covered beneath with the large, thin, imbricated indusia which are orbicular with a narrow sinus, having the margins ragged and sparingly glanduliferous. Stipe short and chaffy.

The fragrant fern grows on high cliffs among the mountains of northern New England. It is reported from scattered stations in northern Maine, from north of the White Mountains and from Sunapee Lake in New Hampshire, and in the Green Mountains south to central Vermont, New Brunswick and to Minnesota. Found also in Alaska and Greenland. This much-coveted fern has a singularly sweet and lasting fragrance, compared by some to strawberries, by others to new-mown hay and sweet brier leaves. We have seen herbarium specimens that were mildly and pleasantly odorous after several years. When growing the fern may be tested "by its fragrance, its stickiness and its beautiful brown curls." Evergreen. Spores ripen the middle of August.

[Illustration: Fragrant Fern. Aspidium fragrans (Mt. Mansfield. Vt.)]

KEY TO THE WOOD FERNS

ASPIDIUM Fronds pinnate, the pinn機innatifid; Blade soft and thin, not evergreen; Lower pinn梼educed to mere lobes New York Fern Lower pinn栢ut slightly reduced; Veins simple.......................Massachusetts Fern Veins forked...............................Marsh Fern Blade rather thick (subcoreaceous) mostly evergreen; Fronds small, narrow, glandular, rock species Fragrant Fern Fronds large, two or more feet high; Lower pinn梿hort, broadly triangular Crested Shield Fern Lower pinn栬onger; Sori close to the margin.... Marginal Shield Fern Sori nearer the midvein; Frond lanceolate....................Male Fern Frond ovate..............Goldie's Shield Fern Fronds twice pinnate with the lower pinnules pinnatifid Boott's Shield Fern Fronds nearly thrice pinnate................Spinulose Shield Fern

[Illustration: Marginal Shield Fern. Aspidium marginale]

THE WOOD FERNS

The ferns of this group, not counting the small fragrant fern, prefer the woods or at least shady places. Although the genus Polstichum represents the true shield ferns, the wood ferns are also thus designated, as their indusia have nearly the shape of small, roundish shields. The old generic name for them all was Asp□um (meaning shield), first published in 1800. For a long time its chief rival was Nephr□m (kidney-like), 1803. Many modern botanists have preferred the earlier name Dry□ris (1763), meaning oak fern, alluding, perhaps, to its forest-loving habits. THEL⌣TERIS, still earlier (1762), may supersede the others.

[Illustration: Marginal Shield Fern. Aspidium marginale (From Woolson's "Ferns," Doubleday, Page & Co.)]

[Illustration: Sori of Marginal Shield Fern]

(1) MARGINAL SHIELD FERN, EVERGREEN WOOD FERN

Asp□um margin&. THEL ᴗTERIS MARGIN

Dry□ris margin□s. Nephr□m margin&l

Fronds from a few inches to three feet long, ovate-oblong, somewhat leathery, smooth, twice pinnate. Pinn桓anceolate, acuminate, broadest just above the base. Pinnules oblong, often slightly falcate, entire or toothed. Fruit-dots large, round, close to the margin. Rocky hillsides in rich woods, rather common throughout our area. The heavy rootstock rises slightly above the ground and is clothed at the crown with shaggy, brown scales. Its rising caudex, often creeping for several inches over bare rocks, suggests the habit of a tree fern. In early spring it sends up a graceful circle of large, handsome, bluish-green blades. The stipes are short and densely chaffy. No other wood fern endures the winter so well. The fronds burdened with snow lop over among the withered leaves and continue green until the new ones shoot up in the spring. It is the most valuable of all the wood ferns for cultivation.

(2) THE MALE FERN

Asp •um F 졀x-mas. THEL ⏝TERIS FIX-MAS

Dry • ris F 졀x-mas. Nephr • m F 졀x-mas

Fronds lanceolate, pinnate, one to three feet high growing in a crown from a shaggy rootstock. Pinn桓anceolate, tapering from base to apex. Pinnules oblong, obtuse, serrate at the apex, obscurely so at the sides, the basal incisely lobed, distant, the upper confluent. Fruit-dots large, nearer the mid vein than the margin, mostly on the lower half of each fertile segment.

The male fern resembles the marginal shield fern in outline, but the fronds are thinner, are not evergreen, and the sori are near the midvein. Its use in medicine is of long standing. Its rootstock produces the well-known f졀x-mas of the pharmacist. This has tonic and astringent properties, but is mainly prescribed as a vermifuge, which is one of the names given to it. In Europe it is regarded as the typical fern, being oftener mentioned and fig-

ured than any other. In rocky woods, Canada, Northfield, Vt., and north-west to the great lakes, also in many parts of the world.

[Illustration: The Male Fern. Aspidium Filix-mas (Vermont)]

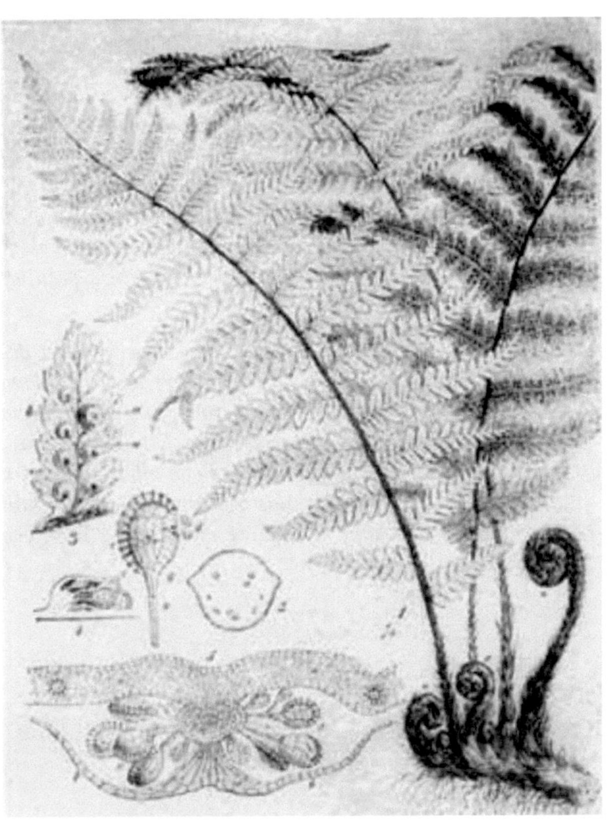

[Illustration: FIG. 33G. Aspidium filix mas 1, Illustration exhibiting general habit; a, young leaves: 2, transverse section of rhizome showing the conducting bundles; a: 3, portion of the leaf bearing sori; a indusium b, sporangia; 4, longitudinal; 5, transverse section of a soris; a, leaf; b, indusium; c, sporangia: 6, a single sporangium; a, stalk; c, annulus; d, spores. (After WOSSIDLO OFFICINAL) From a German print, giving details]

(3) GOLDIE'S FERN

Aspidium Goldiஎ ன். THELYPTERIS GOLDI

A

Dryopteris Goldiana. Nephrodium Goldianum

Fronds two to four feet high and often one foot broad, pinnate, broadly ovate, especially the sterile ones. Pinnæ deeply pinnatifid, broadest in the middle. The divisions (eighteen or twenty pairs) oblong-linear, slightly toothed. Fruit-dots very near the midvein. Indusium large, orbicular, with a deep, narrow sinus. Scales dark brown to nearly black with a peculiar silky lustre.

A magnificent species, the tallest and largest of the wood ferns. It delights in rich woodlands where there is limestone. Its range is from Canada to Kentucky. While not common, there are numerous colonies in New England. It is reported from Fairfield, Me., Spencer and Mt. Toby, Mass., and frequently west of the Connecticut River. We have often admired a large and beautiful colony of it on the west side of Willoughby Mountain in Vermont. It is easily cultivated and adds grace and dignity to a fern garden.

[Illustration: Goldie's Shield Fern. Aspidium Goldianum (Vermont, 1874. C.G. Pringle) (Herbarium of G.E. Davenport)]

[Illustration: Goldie's Fern (From Woolson's "Ferns," Doubleday, Page & Co.)]

(4) THE CRESTED FERN

Aspidium crist ʌ *lm. THEL ᴜTERIS CRIST ¶A*

Dryopteris crist ʌʷ. *Nephrodium crist* ʌ *lm*

Fronds one to two feet long, linear-oblong or lanceolate, pinnate, acute. Pinn 择 two to three inches long, broadest at the base, triangular-oblong, or the lowest triangular. Divisions oblong, obtuse, finely serrate or cut-toothed, those nearest the rachis sometimes separate. Fruit-dots large, round, half way between the midvein and the margin. Indusium smooth, naked, with a shallow sinus.

The short sterile fronds, though spreading out gracefully, are conspicuous only in winter; while the fertile fronds, tall, narrow and erect, are found only in summer.

It is one of our handsomest evergreen ferns and even the large sori, with their dark spore cases and white indusia, are very attractive. The fertile pinn 择ave a way of turning their faces upward toward the apex of the frond for more light. In moist land, Canada to Kentucky.

150

Var. Clintoniæm. Clinton's Wood Fern. Resembles the type, but is in every way larger. Divisions eight to sixteen pairs. Fruit-dots near the midvein, the sides of the sinus often overlapping. South central Maine to New York and westward. "Rare in New England attaining its best development in western sections." (Dodge.) Mt. Toby, Mass., Hanover, N.H. July. Fine for cultivation.

[Illustration: Crested Shield Fern. Aspidium cristatum (Reading, Mass., Kingman)]

[Illustration: The Crested Shield Fern. Aspidium cristatum]

[Illustration: Clinton's Wood Fern. Aspidium cristatum, var. Clintonianum (Gray Herbarium)]

CRESTED MARGINAL FERN

Aspidium cristatum X marginale

Both the crested fern and Clinton's fern appear to hybridize with the marginal shield fern with the result that the upper part of the frond is like marginale and the lower like cristatum, including the veining and texture.

This form was discovered by Raynal Dodge, verified by Margaret Slosson and described by Geo. E. Davenport, who had a small colony under cultivation in his fern garden at Medford, Mass., and to him the writer and other friends are indebted for specimens.]

Found occasionally throughout New England and New Jersey. Other supposed hybrids have been found between the marginal shield and the spinulose fern and its variety intermedium, and with Goldie's fern; also between the crested fern, including Clinton's variety and each of the others mentioned; and, in fact, between almost all pairs of species of the wood ferns, although we do not think they have been positively verified. Still other species of ferns are known to hybridize more or less, as we saw in the case of Scott's spleenwort.

[Illustration: Crested Marginal Fern. A Hybrid. Aspidium Cristatum X marginale (Fernery of Geo. E. Davenport)]

[Illustration: Aspidium cristatum X marginale One of the very best for cultivation]

(5) BOOTT'S SHIELD FERN

Aspidium Boottii. THELYPTERIS BOOTTII

Dryopteris Boottii. Nephrodium Boottii

Fronds one to three feet high, oblong-lanceolate, bipinnate, the upper pinnæ lanceolate, the lower triangular with spinulose teeth. Sori in rows each side of the midvein, one to each tooth and often scattering on the lower pinules. Indusium large, minutely glandular, variable.

This fern has been thought to be a hybrid between the crested and spinulose ferns, but is now regarded as distinct. Like the crested fern its fertile fronds wither in autumn, while its sterile blades remain green throughout the winter. It differs from it, however, by being twice pinnate below, and from the typical spinulose fern by its glandular indusium; but from the intermediate variety it is more difficult to separate it, as that also has indusiate glands. The collector needs to study authentic specimens and have in mind the type, with its rather long, narrow blade as an aid to the verbal description, and even then he will often find it an interesting puzzle. Shaded swamps throughout our area.

[Illustration: Aspidium Boottii]

(6) SPINULOSE SHIELD FERN

Aspidium spinul☐. THEL ‿TERIS SPINULⱫA

Dryopteris spinul · Nephrodium spinul ·

Stipes with a few pale brown deciduous scales. Fronds one to two and one-half feet long, ovate-lanceolate, twice pinnate. Pinn柄blique to the rachis, the lower ones broadly triangular, the upper ones elongated. Pinnules on the inferior side of the pinn柄ften elongated, especially the lower pair, the pinnule nearest the rachis being usually the longest, at least in the lowest pinn殞Pinnules variously cut into spinulose-toothed segments. Indusium smooth, without marginal glands.

The common European type, but in this country far less common than its varieties. They all prefer rich, damp woods, and because of their graceful outline and spiny-toothed lobes are very attractive. They can be transplanted without great difficulty, and the fern garden depends upon them for its most effective lacework.

Var. interm襲um has the scales of the stipe brown with darker center. Fronds ovate-oblong, often tripinnate. Pinn楮preading, oblong-lanceolate. Pinnules pinnately cleft, the oblong lobes spinulose-toothed at the apex. Margin of the indusium denticulate and beset with minute, stalked glands. In woods nearly everywhere--our most common form. Millions of fronds of this variety are gathered in our northern woods, placed in cold storage and sent to florists to be used in decorations.[A] As long as the roots are not disturbed the crop is renewed from year to year, and no great harm seems to result. Canada to Kentucky and westward.

[Illustration: Spinulose Shield Fern. Aspidium spinulosum (Maine, 1877, Herbarium of Geo. E. Davenport)]

[Illustration *medium*]

[Illustration: *Aspidium spinulosum*, var. AMERICANUM]

A tripinnate form of this variety discovered at Concord, Mass., by Hen-
ry   Purdie,   has   been   named   var.   CONCORDI

UM. *It has small, elliptical, denticulate pinnules and a glandular-pubescent indusium.*

Var. *AMERIC*

UM (=dilatatum, syn.). Fronds broader, ovate or triangular-ovate in outline. A more highly developed form of the typical plant, the lower
pinnæ being often very broad, and the fronds tripinnate. Inferior pinnules
on the lower pair of pinnæ conspicuously elongated. A variety preferring
upland woods; northern New England, Greenland to the mountains of
North Carolina, Pennsylvania, Michigan and northward.

THE BLADDER FERNS. Cyst□ris

"Mark ye the ferns that clothe these dripping rocks, Their hair-like stalks, though trembling 'neath the shock Of falling spraydrops, rooted firmly there."

The bladder ferns are a dainty, rock-loving family partial to a limestone soil. (The Greek name cyst□ris means bladder fern, so called in allusion to the hood-shaped indusium.)

(1) THE BULBLET BLADDER FERN

Cyst□ris bulb□ra. F 聚x bulb • ra

Fronds lanceolate, elongated, one to three feet long, twice pinnate. Pinn捋 lanceolate-oblong, pointed, horizontal, the lowest pair longest. Rachis and pinn捋 often bearing bulblets beneath. Pinnules toothed or deeply lobed. Indusium short, truncate on the free side. Stipe short.

[Illustration: Bulblet Bladder Fern. Cystopteris bulbifera (Willoughby, Vt., 1904, G.H.T.)]

[Illustration: Bulblet Bladder Fern. Cystopteris bulbifera]

One of the most graceful and attractive of our native ferns; an object of beauty, whether standing alone or massed with other growths. It is very easily cultivated and one of the best for draping. "We may drape our homes by the yard," says Woolson, "with the most graceful and filmy of our common ferns, the bladder fern." This fern and the maidenhair were introduced into Europe in 1628 by John Tradescant, the first from America.

It delights in shaded ravines and dripping hillsides in limestone districts. While producing spores freely it seems to propagate its species mainly by bulblets, which, falling into a moist soil, at once send out a pair of growing roots, while a tiny frond starts to uncoil from the heart of the bulb. Mt. Toby, Mass., Willoughby Mountain, Vt., calcareous regions in

Maine, and west of the Connecticut River, Newfoundland to Manitoba, Wisconsin and Iowa; south to northern Georgia, Alabama and Arkansas.

(2) THE COMMON BLADDER FERN

Cystopteris fragilis. Filix fragilis

Stipe long and brittle. Fronds oblong-lanceolate, five to twelve inches long, twice pinnate, the pinnæ often pinnatifid or cut-toothed, ovate-lanceolate, decurrent on the winged rachis. Indusium appearing acute at the free end. Very variable in the cutting of the pinnules.

The fragile bladder fern, as it is often called, and which the name fragilis suggests, is the earliest to appear in the spring, and the first to disappear, as by the end of July it has discharged its spores and withered away. Often, however, a new crop springs up by the last of August, as if Nature were renewing her youth. In outline the fragile bladder fern suggests the blunt-lobed Woodsia, but in the latter the pinnæ and pinnules are usually broader and blunter, and its indusium splits into jagged lobes. Rather common in damp, shady places where rocks abound. In one form or another, found nearly throughout the world though only on mountains in the tropics.

[Illustration: Fragile Bladder Fern, Fruited Portion]

[Illustration: Fragile Bladder Fern. Cystopteris fragilis (Wakefield, Mass.)]

KEY TO THE WOODSIAS

Stipes not jointed: Indusium ample, segments broad, frond without hairs. Obtuse Woodsia. Pinnae hispidulous, with white jointed hairs beneath. Rocky Mountain Woodsia. Fronds bright green, pinnae glabrous, oblong. Oregon Woodsia. Fronds dull green, lanceolate, glandular beneath. Cathcart's Woodsia. Stipes obscurely jointed near the base: Fronds more or less chaffy, pinnae oblong to ovate, crowded. Rusty Woodsia. Fronds linear, smooth, pinnae deltoid or orbicular. Smooth Woodsia. Fronds lanceolate, a few white scales beneath; pinnae deltoid-ovate. Alpine Woodsia.

THE WOODSIAS

Small, tufted, pinnately divided ferns. Fruit-dots borne on the back of simply forked, free veins. Indusium fixed beneath the sori, thin and often evanescent, either small and open, or early bursting at the top into irregular pieces or lobes. (Named for James Woods, an English botanist.)

(1) RUSTY WOODSIA. Wo□a ilv鯑is

Fronds oblong-lanceolate, three to ten inches high, rather smooth above, thickly clothed underneath with rusty, bristle-like chaff. Pinnate, the pinn栣rowded, sessile, cut into oblong segments. Fruit-dots near the margin often confluent at maturity. Indusium divided nearly in the center into slender hairs which are curled over the sporangia. Stipes jointed an inch or so above the rootstock.

[Illustration: Rusty Woodsia, Woodsia ilvensis]

The rusty Woodsia is decidedly a rock-loving fern, and often grows on high cliffs exposed to the sun; its rootstock and fronds are covered with silver-white, hair-like scales, especially underneath. These scales turn brown in age, whence the name, rusty. As the short stipes separate at the joints from the rootstock, they leave at the base a thick stubble, which serves to identify the fern. Exposed rocks, Labrador to North Carolina and westward. Rather common in New England. Said to be very abundant on the trap rock hillocks about Little Falls, N.J., where it grows in dense tufts.

(2) NORTHERN WOODSIA. ALPINE WOODSIA

Woodsia alp裂. Woodsia hyperb·

Fronds narrowly lanceolate, two to six inches long, smooth above, somewhat hairy beneath, pinnate. Pinn梅riangular-ovate, obtuse, lobed, the lobes few and nearly entire. Fruit-dots rarely confluent. Indusium as in Woodsia ilvensis.

[Illustration: Details of Northern Woodsia. Woodsia alpina]

Thought by some botanists to be a smooth form of Woodsia ilvensis. It was discovered in the United States by Horace Mann, in 1863, at Willoughby Lake, Vt. Twenty years or more later it was collected by C.H. Peck in the Adirondacks, who supposed it to be Woodsia glab屬la. In 1897 it was rediscovered at Willoughby Lake by C.H. Pringle. New York, Vermont, Maine, and British America. Rare.

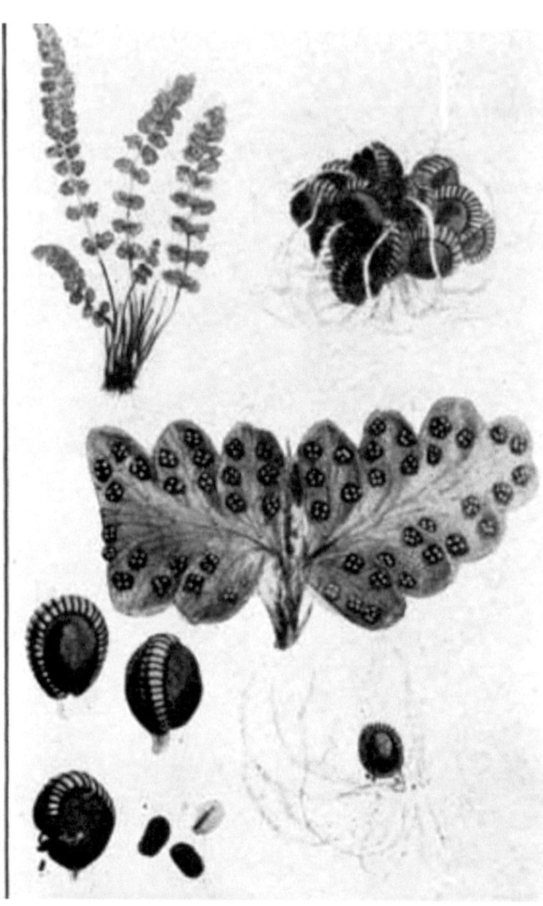

(3) BLUNT-LOBED WOODSIA. Woodsia obtsa

Fronds broadly lanceolate, ten to eighteen inches long, nearly twice pinnate, often minutely glandular. Pinn栲ather remote, triangular-ovate or oblong, pinnately parted into obtuse, oblong, toothed segments. Veins forked. Fruit-dots on or near the margin of the lobes. Indusium conspicuous, at length splitting into several spreading, jagged lobes.

174

[Illustration: Blunt-lobed Woodsia. Woodsia obtusa]

This is our most common species of Woodsia and it has a wider range than the others, extending from Maine and Nova Scotia to Georgia and westward. On rocky banks and cliffs. The sori of this species have a peculiar beauty on account of the star-shaped indusium, as it splits into fragments. Var. angsta is a form with very narrow fronds and pinn殗Highlands, New York. The type grows in Middlesex County, Mass., but is rare.

(4) SMOOTH WOODSIA. *Woodsia glab闘a*

Fronds two to five inches high, very delicate, linear, pinnate. Pinn栲emote at the base, roundish-ovate, very obtuse with a few crenate lobes. Stipes jointed, straw-colored. Hairs of the indusium few and minute.

[Illustration: Smooth Woodsia. Woodsia glabella (Willoughhy Mountain, Vt. G.H.T.)]

On moist, mossy, mostly calcareous rocks, northern New England, Mount Mansfield, Willoughby, and Bakersfield Ledge, Vt., Gorham, N.H., also Newfoundland, New York, and far to the northwest. Not very common. It differs from the alpine species by the absence of scales above the joint. As the name implies, the plant is smooth, except for the chaffy scales at or near the rootstock, which mark all the Woodsias, and many other ferns, and which serve as a protective covering against sudden changes in extremes of heat and cold.

(5) OREGON WOODSIA. *Wo •a oreg •*

Fronds two to ten inches high, smooth, bright green, glandular beneath, narrowly lance-oblong, bipinnatifid. Pinnse triangular-oblong, obtuse, pinnatifid. Segments ovate or oblong, obtuse, crenate, the teeth or margin nearly always reflexed. Indusium minute, concealed beneath the sorus, divided into a few beaded hairs.

Like the obtuse Woodsia this fern has no joint near the base of the stipe, but is much smaller and has several points of difference. Limestone cliffs, Gasp頀eninsula, southern shore of Lake Superior, Colorado, Oregon to the northwest. Its eastern limit is northern Michigan.

(6) ROCKY MOUNTAIN WOODSIA. Wo□a scopul 裂

Fronds six to fifteen inches long [smooth], lanceolate, pinnatifid. Pinn 挥 triangular-ovate, the lowest pair shortened. Under surface of the whole frond hispidulous with minute, white hairs and stalked glands. Indusium hidden beneath the sporangia, consisting mostly of a few hair-like divisions.

In crevices of rocks, mountains of West Virginia, Gasp 預 eninsula, Rocky Mountains, and westward to Oregon and California.

(7) CATHCART'S WOODSIA. Woodsia Cathcarti ·

Fronds eight to twelve inches high, lanceolate, bipinnatifid, finely glandular-puberulent. Pinnse oblong; the lower distant segments oblong, denticulate, separated by wide sinuses.

Rocky river banks, west Michigan to northeast Minnesota.

DENNSTAUTIA. Dicks ·

Fruit-dots small, globular, marginal, each on the apex of a vein or fork. Sporangia borne on an elevated, globular receptacle in a membranous, cup-shaped indusium which is open at the top.

(Named in honor of August Wilhelm Dennstaed.)

HAYSCENTED FERN. BOULDER FERN

DENNSTAUTIA PUNCTILжicULA[A]

Dicks · punctil ·a. Dicks · pilosiscula

[Footnote A: We again remind our readers that the Latin names in small capitals represent the newer nomenclature.]

Fronds one to three feet high, minutely glandular and hairy, ovate-lanceolate, pale green, very thin and mostly bipinnate. Primary pinn 栶 outline like the frond; the secondary, pinnatifid into oblong and obtuse, cut-toothed lobes. Fruit-dots minute, each on a recurved toothlet, usually one at the upper margin of each lobe. Indusium fixed under the sporangia, appearing like a tiny green cup filled with spore cases.

[Illustration: Hayscented Fern. Dennst棩ia punctilobula (Sudbury, Mass. G.E.D.)]

[Illustration: Forked Variety of Hayscented Fern]

[Illustration: Hayscented Fern. Dennst槭ia punctilobula]

While Dennsta餙ia is the approved scientific name of this species, the name Dicks • has come to be used almost as commonly as hay scented fern or boulder fern. It is one of our most graceful and delicate species, its long-tapering outline suggesting the bulblet bladder fern. It delights to cluster around rocks and boulders in upland fields and pastures and in the margin of rocky woods. It is sweet-scented in drying. A fine species for the fernery and one of the most decorative of the entire fern family. The effect of the shimmering fronds, so delicately wrought, flanked by evergreens, is highly artistic. Fine-haired mountain fern, pasture fern, and hairy Dicks • are other names. Canada to Tennessee and westward.

Var. cristata has the fronds more or less forked at the top.

[Illustration: Pinnule and Sori]

[Illustration: Mass of Sensitive Fern]

THE SENSITIVE AND OSTRICH FERNS

Onocl枝/i>. PTERɔIS. Matt贛cia. Struthi • ris

(Last three names applied to Ostrich Fern only.)

It is a question whether the sensitive and ostrich fern should be included in the same genus. They are similar in many respects, but not in all. The sensitive fern has a running rootstock, scattered fronds, and netted veins; while the ostrich fern has an upright rootstock, fronds in crowns, and free veins.

[Illustration: Sensitive Fern. Gradations from Leaf to Fruit. Obtusilobata Form]

(1) SENSITIVE FERN. Onocl衡sens •lis

Fronds one to three feet high, scattered along a creeping rootstock, broadly triangular, deeply pinnatifid, with segments sinuately lobed or nearly entire. Veins reticulated with fine meshes. The fertile fronds shorter, closely bipinnate with the pinnules rolled up into berry-like structures which contain the spore cases. (The name in Greek means a closed vessel, in allusion to the berry-like fertile segments.) The sensitive fern is so called from its being very sensitive to frost. The sterile and fertile fronds are totally unlike, the latter not coming out of the ground until about July, when they appear like rows of small, green grapes or berries, but soon turn dark and remain erect all winter, and often do not discharge their spores until the following spring. The little berry-like structures of the fertile frond represent pinnules, bearing fruit-dots, around which they are closely rolled. As Waters remarks, "Most ferns hold the sori in the open hand, but the sensitive fern grasps them tightly in the clenched fist."

Var. obtusilobata is an abortive form with the fertile segments only partially developed. The illustration shows several intermediate forms.

[Illustration: Sori of Sensitive Fern]

[Illustration: Se
Fronds on one Sto
lection of Mr. and

[Illustration: Sensitive Fern. *Onoclea sensibilis*]

[Illustration: Ostrich Fern. Onoclea Struthiopteris. Fertile Fronds]

(2) OSTRICH FERN

Onoclea struthi□ris. PTERETIS NODULOSA

Struthi□ris Germ • ca. Matteccia struthi • ris

Fronds two to eight feet high, growing in a crown; broadly lanceolate, pinnate, the numerous pinn深eeply pinnatifid, narrowed toward the chan-neled stipe. Fertile fronds shorter, pinnate with margins of the pinn栲evolute into a necklace form containing the sori.

[Illustration: Ostrich Fern. Sterile Fronds (New Hampshire)]

The rootstocks send out slender, underground stolons which bear fronds the next year. Sterile fronds appear throughout the summer, fertile ones in July. Seen from a distance its graceful leaf-crowns resemble those of the cinnamon fern. An intermediate form between the fertile and sterile fronds is sometimes found, as in the sensitive fern. This handsome species thrives under cultivation. For grace and dignity it is unrivaled, and for aggressiveness it is, perhaps, equaled only by the lady fern. For the climax of beauty it should be combined with the maidenhair. The ostrich fern is fairly common in alluvial soil over the United States and Canada.

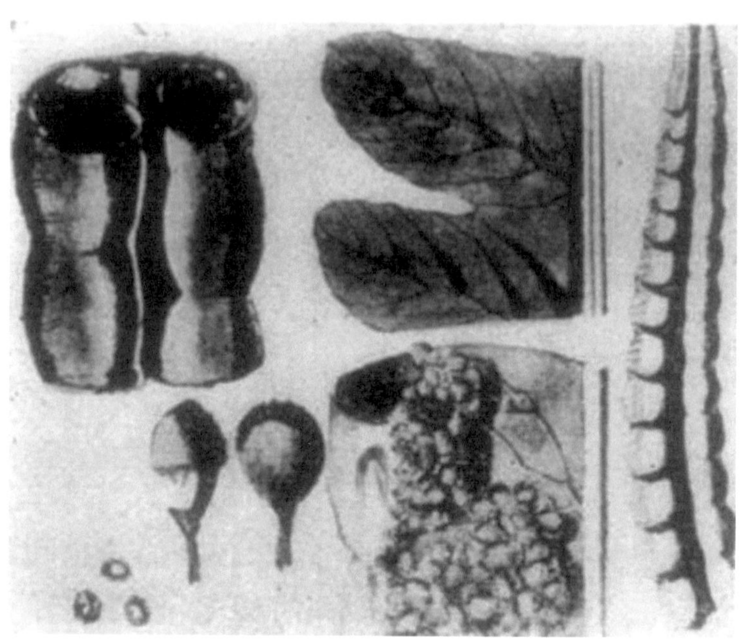

[Illustration: Sori and sporangia of Ostrich Fern]

II

THE FLOWERING FERN FAMILY

OSMUND ᴸEAE

This family is represented in North America by three species, all of which belong to the single genus.

OSMṦDA

The osmundas are tall swamp ferns growing in large crowns from strong, thickened rootstocks; the fruiting portion of the fertile frond much contracted and quite unlike the sterile. Sporangia large, globular, short-stalked, borne on the margin of the divisions and opening into two valves by a longitudinal slit. Ring obscure. (From Osmunder, a name of the god Thor.)

(1) FLOWERING FERN, ROYAL FERN

Osmnda reg • s. Osmunda reg • s, var. SPECT₇ ILIS

Fronds pale green, one to six feet high; sterile part bipinnate, each pinna having numerous pairs of lance-oblong, serrulate pinnules alternate along the midrib. Fruiting panicle of the frond six to twelve inches long, brown when mature and sometimes leafy.

A magnificent fern, universally admired. Well named by the great Linn漈, regalis, royal, indeed, in its type of queenly beauty. The wine-colored stipes of the uncoiling fronds shooting up in early spring, lifting gracefully their pink pinn校nd pretty panicles of bright green spore cases, throw an indescribable charm over the meadows and clothe even the wet, stagnant swamps with beauty nor is the attraction less when the showy fronds expand in summer and the green sporangia are turned to brown. The stout rootstocks are often erect, rising several inches to a foot above the ground, as if in imitation of a tree fern. The poet Wordworth hints at somewhat different origin of the name from that given here.

[Illustration: Royal or Flowering Fern Osmunda regalis]

The royal fern may be transplanted with success if given good soil, sufficient shade and plenty of water. Common in swamps and damp places. Newfoundland to Virginia and northwestward.

[Illustration: Sori of Osmunda regalis (From Waters's "Ferns," Henry Holt & Co.)]

(2) INTERRUPTED FERN. CLAYTON'S FERN

Osmunda Claytoni☐

Fronds pinnate, one to five feet high. Pinnæ cut into oblong, obtuse lobes. Fertile fronds taller than the sterile, having from one to five pairs of intermediate pinnæ contracted and bearing sporangia.

[Illustration: Interrupted Fern. Osmunda Claytoni□]

The fronds have a bluish-green tint; they mature their spores about the last of May. The sterile fronds may be distinguished from those of the cinnamon fern by not having retained, like those, a tuft of wool at the base of each pinna. Besides, in Clayton's fern the fronds are broader, blunter and thinner in texture, and the segments more rounded; the fronds are also more inclined to curve outwards. They turn yellow in the fall, at times "flooding the woods with golden light," but soon smitten by the early frosts they wither and disappear. The interrupted fern is rather common in damp, rocky woods and pastures; Newfoundland to Minnesota, south to North Carolina and Missouri. Although fond of moisture it is easily culti-vated and its graceful outlines make it worthy of a prominent place in the fern garden. Var. dubia has the pinnules of the sterile frond widely separated, and the upper-middle ones much elongated. Southern Vermont.

[Illustration: Interrupted Fern with the Fertile Pinnules Spread Open]

(3) CINNAMON FERN. BRAKES

Osmunda cinnamomea

Fronds one to six feet long, pinnate. Pinnæ lanceolate, pinnatifid with oblong, obtuse divisions. Fertile pinnæ in separate fronds, which are contracted and covered with brown sporangia.

[Illustration: Cinnamon Fern. Leaf Gradations]

[Illustration: Cinnamon Fern. Gradations from Sterile to Fertile Fronds]

[Illustration: Cinnamon Fern, var. frondosa]

Each fertile frond springs up at first outside the sterile ones, but is soon surrounded and overtopped by them and finds itself in the center of a charming circle of green leaves curving gracefully outwards. In a short time, however, it withers and hangs down or falls to the ground. The large, conspicuous clusters of cinnamon ferns give picturesqueness to many a moist, hillside pasture and swampy woodyard. In its crosier stage it is wrapped in wool, which falls away as the fronds expand, but leaves, at the base of each pinna, a tiny tuft, as if to mark its identity.

[Illustration: Cinnamon Fern, var. incisa (Maine)]

Many people in the country call the cinnamon fern the "buckhorn brake," and eat with relish the tender part which they find deep within the crown at the base of the unfolding fronds. This is known as the "heart of Osmund." The fern, itself, with its tall, recurving leaves makes a beautiful ornament for the shady lawn, and like the interrupted fern is easy to cultivate. The spores of all the osmundas are green, and need to germinate quickly or they lose their vitality. Common in low and swampy grounds in eastern North America and South America and Japan. May. Some think it was this species which was coupled with the serpent in the old rhyme,

"Break the first brake you see, Kill the first snake you see, And you will conquer every enemy."

[Illustration: Osmunda cinnamomea, var. glandulosa (From Waters's "Ferns," Henry Holt & Co.)]

Var. frond☐/i> has its fronds partly sterile below and irregularly fertile towards the summit.

Var. inc헝 has the inner pinnules of some of the pinn栭ore or less cut-toothed.

Var. glandul •/i> has glandular hairs on the pinn夊rachis and even the stipes of the sterile frond. This is known only on the coastal plain from Rhode Island to Maryland.

III

CURLY GRASS FAMILY

SCHIZbCE5/div>

CURLY GRASS. Schiz泵pus□a

Small, slender ferns with linear or thready leaves, the sterile, one to two inches high and tortuous or "curled like corkscrews"; fertile fronds longer, three to five inches, and bearing at the top about five pairs of minute, fruited pinn瑒Sporangia large, ovoid, sessile in a double row along the single vein of the narrow divisions of the fertile leaves, and provided with a complete apical ring. (Schiz泰/i>, from a Greek root meaning to split, alluding to the cleft leaves of foreign species.)

[Illustration: Curly Grass. Schiz枷pusilla]

The curly grass is so minute that it is difficult to distinguish it when growing amid its companion plants, the grasses, mosses, sundews, club mosses, etc. The sterile leaves are evergreen. Pine barrens of New Jersey, Grand Lake, Nova Scotia, and in New Brunswick. Several new stations for the curly grass have recently been discovered in the southwest counties of Nova Scotia by the Gray Herbarium expedition, mostly in bogs and hollows of sandy peat or sphagnum.

[Illustration: Sporangia of Curly Grass]

CLIMBING FERN. HARTFORD FERN

Lyg□m palm⌒m

> *"And where upon the meadow's breast The shadow of the thicket lies." BRYANT.*

Fronds slender, climbing or twining, three to five feet long. The lower pinn彛 (frondlets) sterile, roundish, five to seven lobed, distant in pairs with simple veins; the upper fertile, contracted, several times forked, forming a terminal panicle; the ultimate segments crowded, and bearing the sporangia, which are similar to those of curly grass, and fixed to a veinlet by the inner side next the base, one or rarely two covered by each indusium. (From the Greek meaning like a willow twig [pliant], alluding to the flexible stipes.)

[Illustration: Climbing Fern. Lygodium palmatum]

Fifty years ago this beautiful fern was more common than at present. There was a considerable colony in a low, alluvial meadow thicket at North Hadley, Mass., not far from Mt. Toby, where we collected it freely in 1872. Many used to decorate their homes with its handsome sprays, draping it gracefully over mirrors and pictures. It was known locally as the Hartford fern. Greedy spoilers ruthlessly robbed its colonies and it became scarce, at least in the Mt. Toby region. In Connecticut a law was enacted in 1867 for its protection and with good results. But as Mr. C.A. Weatherby states in the American Fern Journal (Vol. II, No. 4), the encroachments of tillage (mainly of tobacco, which likes the same soil), are forcing it from its cherished haunts, thus jeopardizing its survival. Doubtless an aggressive agriculture is in part responsible for its scarcity in the more northern locality. It is still found here and there in New England, New York and New Jersey; also in Kentucky, Tennessee and Florida, but is nowhere common. The fertile portion dies when the spores mature, but the sterile frondlets remain green through the winter. A handsome species for the fernery in the house or out of doors.

IV

ADDER'S TONGUE FAMILY
OPHIOGLOSS ^LE5/i>

Plants more or less fern-like consisting of a stem with a single leaf. In Ophiogl • m the leaf or sterile segment is entire, the veins reticulated and the sporangia in a simple spike. In Botrchium the sterile segment is more or less incised, the veins free, and the sori in a panicle or compound or rarely simple spike. Sporangia naked, opening by a transverse slit. Spores copious, sulphur-yellow.

ADDER'S TONGUE. Ophiogl • m vulg⌒⋃m

Rootstock erect, fleshy. Stem simple, two to ten inches high, bearing one smooth, entire leaf about midway, and a terminal spike embracing the sporangia, coherent in two ranks on its edges. (Generic name from the Greek meaning the tongue of a snake, in allusion to the narrow spike of the sporangia.)

In moist meadows or rarely on dry slopes. "Overlooked rather than rare." New England states and in general widely distributed. July. Often grows in company with the ragged orchis. The ancient ointment known as "adder's speare ointment" had the adder's tongue leaves as a chief ingredient, and is said to be still used for wounds in English villages.

> *"For them that are with newts or snakes or adders stung, He seeketh out a herb that's called adder's tongue."*

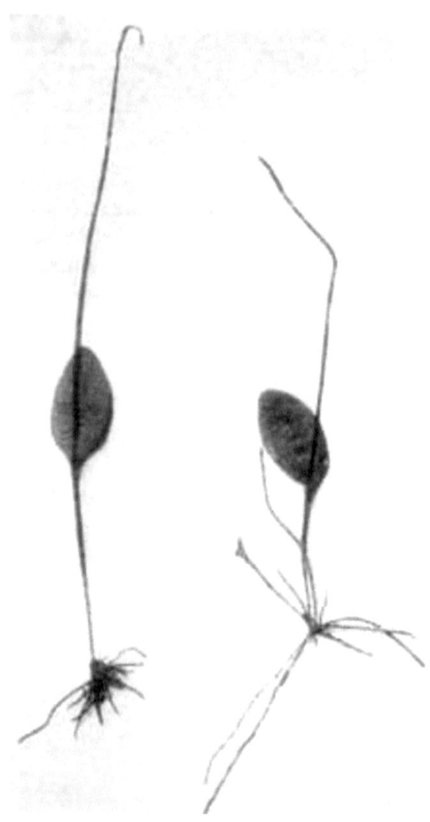

[Illustration: Adder's Tongue. Ophioglossum vulgatum]

Var. minus, smaller; fronds often in pairs. The sterile segment yellowish-green, attached usually much below the middle of the plant. Sandy ground, New Hampshire to New Jersey.

Var. Engelm • i. (Given specific rank in Gray.) Has the sterile segment thicker and cuspidate, the stipe slender and the secondary veins forming a fine network within the meshes of the principal ones. Virginia and westward.

Var. aren • um. (From the Latin, ar i̧f, meaning sand, being found in a sandy soil.) Probably a depauperate form of Ophi •ssum vulg Ω m and about half as large. A colony of these ferns was discovered growing in poor soil at Holly Beach, New Jersey.

(Botrchium) Plant large, fruiting in June, sterile part much divided: Rattlesnake Fern. Plant smaller: Fruiting in autumn, sterile part long-stalked, triangular. Common Grape Fern. Fruiting in summer: Plant fleshy, sterile part mostly with lunate segments. Moonwort. Plant less fleshy, segments not lunate: Sterile part short-stalked above the middle of the stem. Matricary Fern. Sterile part stalked usually below middle of stem. Little Grape Fern. Sterile part sessile near the top of the stem. Lance-leaved Grape Fern.

GRAPE FERNS

Botrchium

Rootstock very short, erect with clustered fleshy roots; the base of the sheathed stalk containing the bud for the next year's frond. Fertile frond one to three pinnate, the contracted divisions bearing a double row of sessile, naked, globular sporangia, opening transversely into two valves. Sterile segment of the frond ternately or pinnately divided or compound. Veins free. Spores copious, sulphur yellow. (The name in Greek means a cluster of grapes, alluding to the grape-like clusters of the sporangia.)

(1) MOONWORT. Botrchium Lun□a

Very fleshy, three to ten inches high, sterile segment subsessile, borne near the middle of the plant, oblong, simple pinnate with three to eight pairs of lunate or fan-shaped divisions, obtusely crenate, the veins repeatedly forking; fertile segment panicled, two to three pinnate.

[Illustration: Moonwort Botrychium Lunaria]

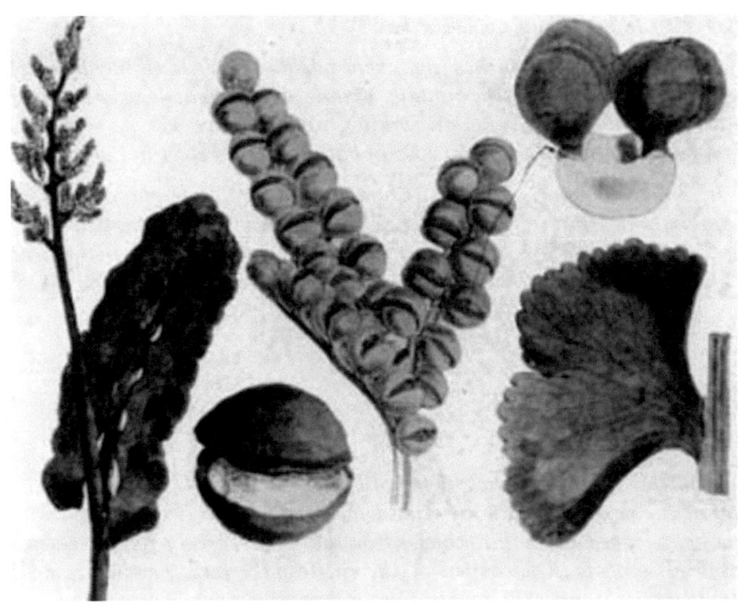

[Illustration: Moonwort. Botrychium Lunaria. Details]

The moonwort was formerly associated with many superstitions and was reputed to open all locks at a mere touch, and to unshoe all horses that trod upon it. "Unshoe the horse" was one of the names given to it by the country people.

> *"Horses that feeding on the grassy hills, Tread upon moonwort with their hollow heels, Though lately shod, at night go barefoot home Their maister musing where their shoes be gone."*

In dry pastures, Lake Superior and northward, but rare in the United States. Willoughby, Vt., where the author found a single plant in 1904, and St. Johnsbury, Vt. Also New York, Michigan and westward.

In England said to be local rather than rare. Sometimes called Lunary.

> *"Then sprinkled she the juice of rue With nine drops of the midnight dew From Lunary distilling." DRAYTON.*

(2) LITTLE GRAPE FERN. Botrychium s☐lex

Fronds two to four inches high, very variable. Sterile segment short-petioled, usually near the middle, simple and roundish or pinnately three to seven lobed. Veins all forking from the base. Fertile segments simple or one to two pinnate, apex of both segments erect in the bud.

In moist woods and fields, Canada to Maryland and westward; Conway and Plainfield, Mass., Berlin and Litchfield, Conn. Rare. According to Pringle it is "abundantly scattered over Vermont, its habitat usually poor soil, especially knolls of hill pastures." May or June.

(3) LANCE-LEAVED GRAPE FERN

Botrychium lanceol☐m

BOTR☐HIUM ANGUSTISEGMΥTUM

Frond two to nine inches high, both sterile and fertile segments at the top of the common stalk. Sterile segment triangular, twice pinnatifid, the acute lobes lanceolate, incised or toothed, scarcely fleshy, resembling a very small specimen of the rattlesnake fern. Fertile segment slightly overtopping the sterile, two to three pinnate and spreading.

One of the constant companions of the rattlesnake fern. New England to Lake Superior. July.

[Illustration: Little Grape Fern Botrychium simplex]

[Illustration: Lance-leaved Grape Fern Botrychium lanceolatum Botrychium angustisegmentum]

(4) MATRICARY FERN

Botrychium ram□. Botrychium matricari□lium

Fronds small, one to twelve inches high. Sterile segment above the middle, usually much divided. Fertile segment twice or thrice pinnate. Apex of both segments turned down in the bud, the sterile overtopping and clasping the fertile one.

214

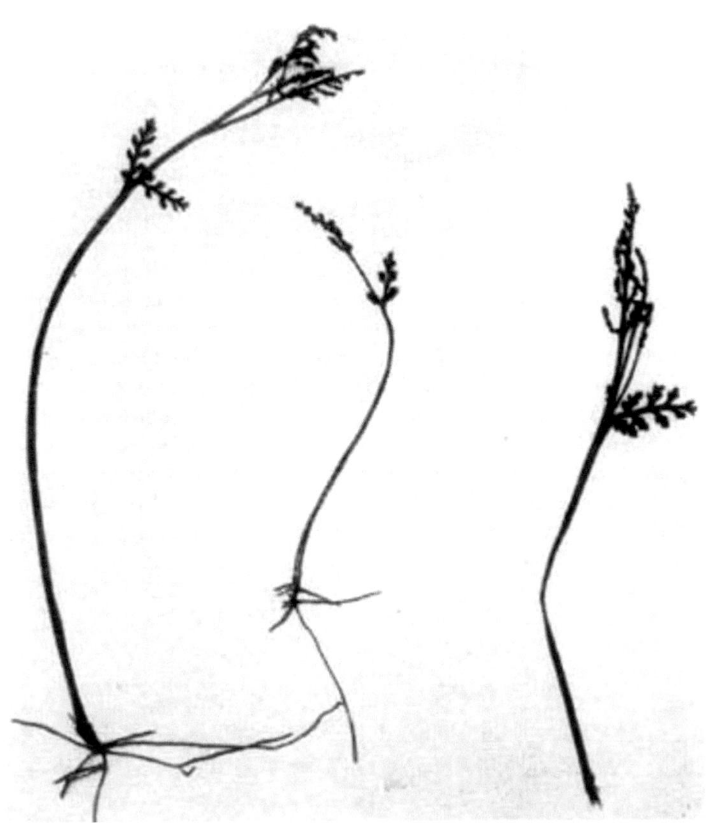

[Illustration: The Matricary Fern Botrychium ramosum]

The matricary fern differs from the preceding in ripening its spores about a month earlier, in having its sterile frond stalked, besides being a taller and fleshier plant. It may also be noted that in the lance-leaved species the midveins of the larger lobes are continuous, running to the tip; whereas in the matricary fern the midveins fork repeatedly and are soon indistinguishable from the veinlets. The two are apt to grow near each other, with the rattlesnake fern as a near neighbor. June.

NOTE. In 1897 A.A. Eaton discovered certain Botrychia in a sphagnum swamp in New Hampshire, to which he gave the specific name of Botrychium tenebr☐. The plants were very small, not averaging above two or three inches high, with the sterile blade sessile or slightly stalked. Many botanists prefer to place this fern as a variety of the matricary, but others regard it as a form of Botrychium s□lex. Borders of maple swamps, Vermont, New Hampshire, Massachusetts, and New York.

(5) COMMON GRAPE FERN

Botrychium obl 창um. Botrychium tern ᘒm, var. obl 창um

BOTRYCHIUM DISSBTUM, *var.* OBLUUM

Rootstock short, its base including the buds of succeeding years. Fronds two to twelve inches or more high. Leafy or sterile segment triangular, ternate, long-petioled, springing from near the base of the plant, and spreading horizontally. From the main leafstock grow several pairs of stalked pinn𝄞with the divisions ovate-oblong, acutish, crenate-serrulate, obliquely cordate or subcordate. Fertile segment taller, erect, about three times pinnate, maturing its fruit in autumn. Occasionally two or three fertile spikes grow on the same plant. In vernation the apex of each segment is bent down with a slight curve inward.

[Illustration: Common Grape Fern. Botrychium obliquum]

New England to Virginia, westward to Minnesota and southward.

Botrychium obliquum, var. dissectum. Similar to the type, but with the divisions very finely dissected or incisely many-toothed, the most beautiful of all the grape ferns. There is considerable variety in the cutting of the fronds. Maine to Florida and westward.

Botrychium obl창um, var. oneid鯵e. Ultimate segments oblong, rounded at the apex, crenulate-serrate, less divided than any of the others and, perhaps, less common. Vermont to Central New York.

Botrychium obl창um, var. elongΩJm. Divisions lanceolate, elongated, acute.

[Illustration: Botrychium obliquum var. oneidense]

Note: A Botrychium not uncommon in Georgia and Alabama, named by Swartz B. lunarioides, deserves careful study. It is known as the "Southern Botrychium."

[Illustration: Botrychium obliquum, var. dissectum]

(6) TERNATE GRAPE FERN

Botrchium tern☊m, var. interm☙um

Botrchium obl☖um, var. interm☙um

Leaf more divided than in obl☖um and the numerous segments not so long and pointed, but large, fleshy, ovate or obovate (including var. austr☙), crenulate, and more or less toothed.

Sandy soil, pastures and open woods. More northerly in its range-- New England and New York. Var. rutaef·m. More slender, rarely over six or seven inches high; sterile segment about two inches broad, its

*divisions few, broadly ovate, the lowest sublunate. The first variety
passes insensibly into the second.*

[Illustration: Ternate Grape Fern Botrychium ternatum var. intermedium (Reduced)]

[Illustration: Ternate Grape Fern Botrychium ternatum var. intermedium (Two stocks, reduced)]

(7) RATTLESNAKE FERN. Botrychium virginianum

Fronds six inches to two feet high. Sterile segment sessile above the middle of the plant, broadly triangular, thin, membranaceous, ternate. Pinnules lanceolate, deeply pinnatifid; ultimate segments oblong or lanceolate and scarcely or not at all spatulate. Fertile part long-stalked, two to three pinnate, its ultimate segments narrow arid thick, nearly opaque in dried specimens. Mature sporangia varying from dark yellow-brown to almost black. Open sporangia close again and are flattened or of a lenticular form. In rich, deciduous woods, rather common and widely distributed.

[Illustration: Rattlesnake Fern. Botrychium virginianum (From Waters's "Ferns," Henry Holt & Co.)]

Prince Edward Island, Minnesota, south to Florida and Texas, and north to Newfoundland and Labrador.

Var. gr□lis. A form much reduced in size.

Var.   LAURENTI

UM. *A conspicuous variety having thick and heavy sterile fronds less finely divided than the type, with the segments crowded to overlapping. Pinnules shorter than the type, tending to be ovate, outer segments strongly spatulate. Fertile spike relatively short and stout, strongly paniculate when well developed. Ultimate segments flat, folaceous, one mm. wide. Mostly confined to the limestone district near the Gulf of St. Lawrence, Labrador, Newfoundland, Quebec, Maine, and Michigan.*

Var. INTERMEDIUM. Segments of sterile fronds ultimately much spatulate, previously ovate, not overlapping. Segments of fertile fronds ultimately narrowly flattened. (For this and the other varieties see Rhodora of September, 1919.) Nova Scotia, Maine, Vermont, Massachusetts, Connecticut, northern New York, Illinois, and Missouri.

Var. EUROP/UM. Fertile frond less finely dissected than in type. Ultimate segments more obtuse than in type; has but very slight tendency towards the spatulate form of the two previous varieties. Pinnules lanceolate, strongly decurrent so that the pinn校re merely pinnatifid. In coniferous forests of Canada, and confined to calcareous regions. Quebec, New Brunswick, New Hampshire, Vermont, New York, Ontario, Montana, and British Columbia. Said to be rare even in Europe.

V

THE FILMY FERN FAMILY
HYMENOPHYLL ᴸE5/i>

The filmy ferns are small, delicate plants with membranaceous, fine-ly dissected fronds from slender, creeping rootstocks. Sporangia sessile on a bristle-like receptacle. There are about one hundred species, mostly tropical, only one of which grows as far north as Kentucky.

[Illustration: Filmy Fern Trichomanes Boschianum (From Waters' "Ferns", Henry Holt & Co.)]

Trich☐es Boschiவ☐m. Trich☐es r • cans

Rootstocks creeping, filiform, stipes ascending, one to three inches long, thin, very delicate, pellucid, much divided, oblong-lanceolate, bipinnatifid. Rachis narrowly winged. Sporangia clustered around the slender bristle, which is the prolongation of a vein, and surrounded by a vase-like, slightly two-lipped involucre.

On moist, dripping sandstone cliffs, Kentucky to Alabama. Often called the "Killarney fern," as it grows about the lakes of Killarney in Ireland.

[Illustration: Fruiting Pinnules of Filmy Fern (From Waters's "Ferns." Henry Holt & Co.)]

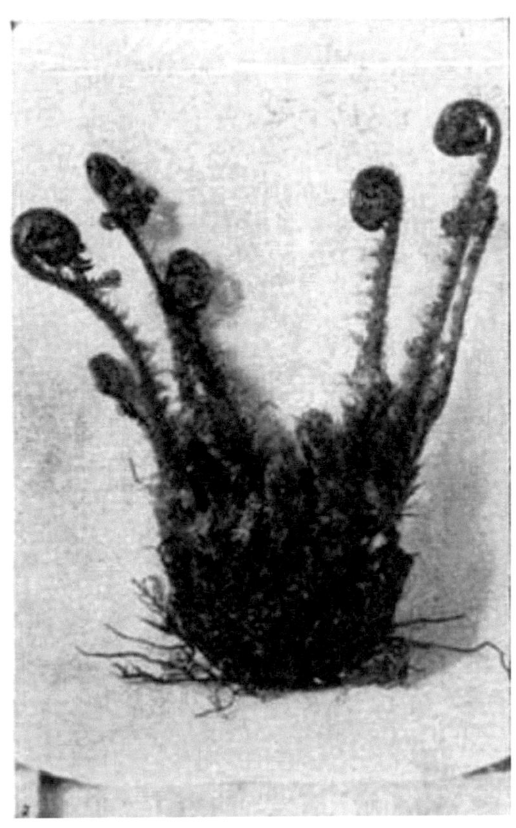

[Illustration: Ostrich Fern]

[Illustration: Cinnamon Fern]

228

[Illustration: Marginal Shield Fern]

[Illustration: Lady Fern Crosiers]

[Illustration: Fiddleheads or Crosiers of Christmas Fern]

BIOGRAPHICAL SKETCHES

[The works of these authors are listed under "Fern Literature" in the following pages.]

EATON, DANIEL CADY. Born at Gratiot, Mich., September 12, 1834. His grandfather was Amos Eaton, noted botanist and author. Studied botany under his friend, Prof. Asa Gray, who had studied with Prof. John Torrey, who in turn was a pupil of Amos Eaton. Daniel C. was professor of botany in Yale College, for more than thirty years. A man of graceful and winsome personality, an authority on ferns, and widely known by his writings. His masterpiece was "The Ferns of North America" in two large, quarto volumes, beautifully illustrated. He died June 29, 1895.

CLUTE, WILLARD NELSON. Born at Painted Post, N.Y., February 26, 1869. Education informal; common schools, university lectures and private study. Manifested early a keen interest in birds and flowers. Was founder and first president of the American Fern Society. Collected in Jamaica more than three hundred species of ferns. Has written extensively on the ferns and their allies, besides publishing several standard volumes. His great distinction is in founding and editing the Fern Bulletin through its twenty volumes, when he combined this publication with The American Botanist, which is now on its twenty-eighth volume, the whole a prodigious achievement of great scientific value.

[Illustration: Noted Writers on Ferns W.N. CLUTE, D.C. EATON, F.T. PARSONS, G. DAVENPORT, J. WILLIAMSON, L.M. UNDERWOOD, W.R MAXON, A.A. EATON, C.E. WATERS, R. DODGE]

UNDERWOOD, LUCIUS MARCUS. *Born at New Woodstock, N.Y., October 26, 1853. Spent early life on a farm. Was graduated from Syracuse University in 1877. After teaching several years in his alma mater and elsewhere, he became Professor of Botany in Columbia University. He contributed numerous articles to the Torrey Bulletin, Fern Bulletin, and other scientific journals. His scholarly book, "Our Native Ferns and Their Allies," continued unexcelled through six editions. He died November 16, 1907.*

DAVENPORT, GEO. EDWARD. *Born in Boston, August 3, 1833. A promoter and officer of the Middlesex Institute. An accurate and dili-*

gent student of the ferns, his numerous articles were published in the Fern Bulletin, in the Torrey Bulletin, Rhodora, and in separate monographs. He was a leading authority on the pteridophyta, and collected a large and choice herbarium of the native ferns, which he donated to the Massachusetts Horticultural Society. By his gentle manners and kindly spirit he won many friends, all of whom were proud to recognize his distinguished ability. He cultivated many of our rare native ferns in his Fellsway home, at Medford, Mass., and freely gave specimens to his friends. He died suddenly of heart failure, November 29, 1907.

WATERS, CAMPBELL EASTER. Born in Baltimore County, Md., September 14, 1872. Was graduated at Johns Hopkins University in 1895. Ph.D. in 1899. Was for a time a close student of ferns, and issued his notable book, "Ferns," in 1903, containing his "Analytical Key Based on the Stipes." A chemist by profession, he has pursued that branch of science for the last eighteen years. His address is Bureau of Standards, Washington, D.C.

MAXON, WILLIAM RALPH. Born at Oneida, N.Y., February 27, 1877. Was graduated at Syracuse University in 1898. Began as aid in cryptogamic botany, United States National Herbarium, 1899, and is now associate curator of the same. Has specialized in scientific work on the pteridophyta, distinguishing himself by the excellence as well as by the large number of his publications, the more important being "Studies of Tropical American Ferns," Nos. 1 to 6. The Fern Bulletin, Torrey Bulletin, American Fern Journal, Fernwort Papers, et al., have profited from his expert and up-to-date knowledge. He is president of the American Fern Society.

PARSONS, FRANCES THEODORA. Born in New York, December 5, 1861. N<#/i> Smith. Married Commander William Starr Dana of the United States Navy, who was lost at sea. As Mrs. Dana, she published, "How to Know the Wild Flowers," in 1893, and within ten years more than seventy thousand copies of the book had been sold. "According to Season" appeared in 1894. In February, 1896, she married Prof. James Russell Parsons, treasurer of the University of the State of New York. In 1899 she published, "How to Know the Ferns." She combined a thorough knowledge of her subject with an easy and graceful style.

DODGE, RAYNAL. Born at Newburyport, Mass., September 9, 1844. Civil War veteran. Wounded at Port Hudson, June 28, 1863. A machinist by trade. A careful observer and student of nature, he discovered Aspidium simulatum at Follymill, Seabrook, N.H., in 1880. (Whittier's "My Playmate," verse 9.) He discovered also the hybrid Aspidium cris-

tatum ₂Marginale. He published his little book, "Ferns and Fern Allies of New England," in 1896. Died October 20, 1918.

EATON, ALVAH AUGUSTUS. Born at Seabrook, N.H., November 20, 1865. Studied at the Putnam School in Newburyport, but was largely self-educated. He took up teaching for several years, spending three years in California. Returning East, he became a florist and began to write for various fern journals, giving special attention to the fern allies. He prepared the genera Equisetum and Isoetes for the seventh edition of "Gray's Manual." He proved the keenness of his observing powers by discovering several ferns new to the United States. Died at his home in North Easton, Mass., September 29, 1908.

WILLIAMSON, JOHN. Born in Abernathy, Scotland, about the year 1838. He came to Louisville, Ky., to live in 1866. A wood-carver by trade, he could work skillfully in wood or metal, and after a time established a brass foundry. His friend, George E. Davenport, writes of him: "He caught as by some divine gift or inspiration the innermost life and feelings of the wild flowers and ferns, and his marvelously accurate needle transfixed them with revivifying power on paper or metal." His "Ferns of Kentucky," issued in 1878, was the first handbook on ferns published in the United States. He died June 17, 1884, in the mountains of West Virginia, whither he had gone for his health.

FERN LITERATURE

AMERICAN FERN JOURNAL. 1910. The American Fern Society. (Annual subscription, $1.25.)

BELAIRS, NONA. Hardy Ferns. Smith, Elder and Co. London, 1865.

BRITISH FERN GAZETTE.

BRITTEN, JAMES. European Ferns. Colored Plates. Cassell & Co. London. Quarto.

BUTTERS, F.K. Athyrium. Study of the American Lady Ferns. Rhodora, September, 1917.

CAMPBELL, D.H. Structure and Development of the Mosses and Ferns. Macmillan & Co. 1905. Ed. 2.

CLUTE, WILLARD N. Our Ferns in Their Haunts. Frederick A. Stokes Co. New York, 1901.

Fern Collector's Guide. Frederick A. Stokes Co. New York, 1902.

The Fern Allies. Frederick A. Stokes Co. New York, 1905.

The Fern Bulletin. Founder and Editor. 20 vols. 1893-1912.

Combined with The American Botanist. Joliet, Ill. 1912.

CONARD, HENRY S. Structure and History of Hayscented Fern. Washington, 1908.

COOK, M.C. Fern-book for Everybody. E. Warne & Co. London.

DAVENPORT, GEO.E. Catalog of Davenport Herbarium, Massachusetts Horticultural Society. 1879. Numerous Monographs and Notes on New England ferns in Torrey Bulletin, Fern Bulletin, and Rhodora. The following monographs are in single booklets by Massachusetts Horticultural Society. Aspidium cristatum ɔmarginale, Aspidium simulatum, Aspidium spinulosum and its Varieties, Botrychium ternatum and its Varieties, Notes on Botrychium simplex.

DODGE, RAYNAL. The Ferns and Fern Allies of New England--very small volume, now out of print. W.N. Clute & Co. 1904.

DRUERY, CHARLES T. British Ferns and Their Varieties. Routledge & Son. London.

EASTMAN, HELEN. New England Ferns and Their Common Allies. Houghton Mifflin & Co. Boston, 1904. Out of print.

EATON, DANIEL C. The Ferns of North America. 2 vols. 1879-80. S.E. Cassino, Salem. Drawings by J.H. Emerton and C.E. Faxon.

EATON, A.A. Specialist in Fern Allies. Prepared Equisetum and Isoetes for Gray's Manual, 7th ed. 1908.

GILBERT, BENJ.D. List of North American Pteridophytes. 1901. Utica, N.Y.

HERVEY, ALPHAEUS B. Wayside Flowers and Ferns. Page & Co. Boston, 1899.

HEMSLEY, ALFRED. Book of Fern Culture. John Lane. London, 1908.

HIBBARD, SHIRLEY. The Fern Garden. Groombridge & Sons. 5 Paternoster Row, London. 1869.

HOOKER, SIR W.J. Genera Filicum. Large 8vo. London, 1842. Contains fine plates which include all American genera. Costs about $25.

Species Filicum. 5 vols. 8vo. London, 1846-64. Vol. II contains seventeen and Vol. Ill contains two plates of American ferns with descriptions of more species. Cost about $50.

HOOKER, SIR W.J., & BAKER. Synopsis Filicum 2d ed. 1874. 8vo. Describes all ferns then known, including the American species. Has also figures illustrating each genus. Costs about $10.

LOWE, EDWARD J. Ferns British and Exotic. 9 vols. 8vo. Bell & Daldy. London, 1868. 550 plates, some very poor. Some American ferns are represented. "The descriptions," says John Robinson, "are worthless, and the synonymy is often incorrect."

MAXON, WILLIAM R. A List of Ferns and Fern Allies of North America, north of Mexico, etc. National Museum, 23:619-651. 1901.

Numerous Monographs and Notes on American Ferns in current magazines.

Studies of Tropical American Ferns. United States National Herbarium, 17:541+.

Pteridophyta (excepting Equisitaceæ and Isoetaceæ of the northern United States, Canada and the British Possessions. In Britton and Brown, Illustrated Flora, etc., ed. 2, pp. 1-54. 1913. New York.

MEEHAN, THOMAS. Native Flowers and Ferns of the United States. Boston, 1878-9.

MOORE, THOMAS. Nature-printed British Ferns. 2 vols. London, 1859.

PARSONS, FRANCES T. How to Know the Ferns. Charles Scribner's Sons. New York, 1899.

PRATT, ANNE. The Ferns of Great Britain and Their Allies. F. Warne & Co. London. No date.

REDFIELD, JOHN. Geographical Distribution of the Ferns of North America. Torrey Bulletin, VI, 1-7. (1875).

RHODORA. Journal of the New England Botanical Club. January, 1899, to date.

ROBINSON, JOHN. Ferns in Their Homes and Ours. S.E. Cassino. Salem, 1878. Out of print.

SACHS, JULIUS. Text Book of Botany. (Translated.) Macmillan & Co. London. 8vo.

SLOSSON, MARGARET. *How Ferns Grow.* Henry Holt & Co. New York. 1906. Out of print.

SMALL, JOHN K. *Ferns of Tropical Florida.* New York, 1918.

SMITH, JOHN. *Historia Filicum.* London, 1875. Amply illustrated, reliable.

STEP, EDWARD. *Wayside and Woodland Ferns.* F. Warne & Co. London, 1908.

TIDESTROM, IVAR. *Elysium Marianum.* Washington, D.C.

UNDERWOOD, LUCIEN M. *Our Native Ferns and Their Allies.* Henry Holt & Co. Edition 6. 1900. Valuable. Out of print.

WATERS, CAMPBELL E. *Ferns.* Henry Holt & Co. 1903. Out of print. Scarce.

WEATHERBY, C.A. *Changes in the Nomenclature of the Gray's Manual of Ferns. Important article in the Rhodora of October, 1919.*

WILLIAMSON, JOHN. *Ferns of Kentucky.* J.P. Morton & Co. Louisville, Ky. 1878.

Fern Etchings. J.P. Morton & Co. 1879. Both out of print.

WOOLSON, GRACE A. *Ferns and How to Grow Them.* Doubleday, Page & Co. New York, 1909.

WRIGHT, MABEL O. *Flowers and Ferns in Their Haunts.* Macmillan & Co. New York, 1901.

[Illustration:

"Fringing the stream at every turn, Swung low the waving fronds of fern." WHITTIER.]

TIMES OF THE FRUITING OF FERNS

"Ah! well I mind the calendar Faithful through a thousand years Of the painted race of flowers."--EMERSON.

Compiled from Dodge's "Ferns and Fern Allies of New England"

May 25. Little Grape Fern. Interrupted Fern. May 30. Cinnamon Fern. June 5. Ostrich Fern. June 10. Frondosa variety of Cinnamon Fern. June 15. Matricary Grape Fern. June 20. Royal Fern. Interrupted Fern. June 25. Rattlesnake Fern. June 30. Oak Fern. Spinulose Wood Fern and Varieties. July 5. Fragile Bladder Fern. Christmas Fern. July 10. Long Beech Fern. Crested Shield Fern. Boott's Shield Fern. July 15. Moonwort. Virginia Chain Fern.

Adder's Tongue. Crested Marginal Shield Fern. July 20. Slender Cliff Brake. Blunt-Lobed Woodsia. July 25. Purple Cliff Brake. Bulblet Bladder Fern. Mountain Spleen wort. July 30. Goldie's Shield Fern. Marginal Shield Fern. Clinton's Wood Fern. August 5. Wall Rue. Walking Fern. Lady Fern. August 10. Alpine Woodsia. Smooth Woodsia. Common Polypody. Maidenhair Fern. Fragrant Shield Fern. Scott's Spleenwort. Braun's Holly Fern. August 15. Rusty Woodsia. Silvery Spleen wort. Lance-leaved Grape Fern. August 20. Ebony and Maidenhair Spleenworts. Hayscented Fern. New York Fern. August 25. Broad Beech Fern. August 30. Marsh Fern. September 5. Bracken or Brake. September 10. Climbing Fern. Narrow-leaved Spleenwort. September 15. Massachusetts Fern. Green Spleenwort. Sensitive Fern. Ternate Grape Fern. September 30. Narrow-leaved Chain Fern.

ACUMINATE. Gradually tapering to a point. ACULEATE. Prickly. Beset with prickles. ACUTE. Sharp pointed, but not tapering. ADVENTIOUS. Irregular, incidental. Growing out of the usual or normal position. ANASTOMOSING. Connected by cross veins and forming a network as in the Sensitive ferns.

NULUS. A jointed, elastic ring surrounding the spore cases in most ferns. ANTHERIA. The male organs on a prothallium. APEX The top or pointed end of leaf or frond. (plu. APICES). ARCHEGPIA. The female organs on a prothallium. ARôLA. A space formed by intersecting veins; a mesh. AURICLE. An ear-shaped lobe at the base. ARTULATE. Jointed; having a joint or node. AXIL. The angle formed by a leaf or branch with the stem. BI (Latin, Two, twice, doubly. bis, twice). BLADE. The expand-ed, leafy portion of a frond. BULBLET. A small bulb, borne on a leaf or in its axil. CAUDATE. With a slender, tail-like append-age. CAUDEX. A trunk or stock of a plant; especially of a tree fern. CHAFF. Thin, dry scales of a yellowish-brown color. CHLFOPHYLL. The green coloring matter of plants. CĬATE. Fringed with fine hairs. CCINATE. Coiled downward from the apex, as in the young fronds of a fern. CLAVATE. Club-shaped. COMPOUND. Divided into two or more parts. CONFLUENT. Blended together. CORDATE. Heart-shaped. CRENATE. Scal-loped with rounded teeth; said of margins. CRₑIER. An uncoiling frond. CŌEATE. Wedge-shaped. CⱼPIDATE. Hard pointed, tipped with a cusp. DECIDUOUS. Falling away when done growing--not evergreen. DECOMPOUND. More than once com-pounded or divided. DECURRENT. Running down the stem be-low the point of insertion, as the bases of some pinn�894DE-CUMBENT. Not erect; trailing, bending along the ground, but with the apex ascending. DEFLEXED. Bent or turned abruptly downward. DENTATE. Toothed. Having the teeth of a margin directed outward. DICHÆOMOUS. Forking regularly in pairs. DIMÄPHOUS. Of two forms; said of ferns whose fertile fronds are unlike the sterile. EMĬGINATE. Notched at the apex. EN-TIRE. Without divisions, lobes, or teeth. FALCATE. Scythe-shaped, slightly curved upward. FERTILE. Bearing spores. FĬFORM. Thread-like; long, slender, and terete. FILMY. Having a thin membrane; gauzy; said of the filmy fern fronds. FLABEL-LATE. Fan-shaped; broad and rounded at the summit and nar-row at the base. FROND. A fern leaf or blade; may include both stipe and blade, or only the latter--called also lamina. GLA-BROUS. Smooth; not rough or hairy. GLAND. A small secreting organ, globular or pear-shaped; it is often stalked. GLAUCOUS. Covered with a fine bloom, bluish-white and powdery, in ap-pearance like a plum. HASTATE. Like an arrowhead with the lobes spreading. IMBRICATE. Overlapping, like shingles on a roof. INCED. Cut irregularly into sharp lobes. INDŌIUM. The thin membrane covering the sori in some ferns. INVOLUCRE. In ferns, an indusium; in filmy ferns, cup-shaped growths encircling the sporangia. L INA. A blade; the leafy portion of a fern. LACIATE. Slashed; cut into narrow, irregular lobes. LANCEO-LATE. Lance-shaped; broadest above the base and tapering to the apex. LOBE. A small rounded segment of a frond. MIDRIB. The main rib or vein of a segment, pinnule, pinna, or frond; a midvein. M RONATE. Ending abruptly in a short, sharp point. OBLONG. From two to four times longer than broad and with sides nearly parallel. OBTUSE. Blunt or rounded at the end. OES. A Greek ending, meaning like, or like to, as polypodioides--like to a polypody. SPHERE. The egg-cell in fern reproduction--becoming the o · · re when fertilized. OVATE. Egg-shaped with the broader end downward. PALMATE. Having lobes radiating like the fingers of a hand. PANICLE. A loose compound cluster of flowers or sporangia with irregular stems. PEDICEL. A tiny stalk, especially the stalk of the sporangia. PELLUCID. Clear, transparent. PERSISTENT. Remaining on the plant for a long time, as leaves through the winter. PIOLE. The same as stalk or stipe. PINNA. One of the primary divisions of a frond. PINNATE. Feather-like; with the divisions of the frond extending fully to the rachis. PINN IFID. Having the divisions of the frond extend halfway or more to the rachis or mid vein. PIN-NULE. A secondary pinna. In a bipinnate frond one of the smaller divisions extending to the secondary midvein. PROCUM-BENT. Lying on the ground. PROTH

LIUM. (Or proth ⋅ us.) A delicate, cellular, leaf-like structure produced from a fern spore, and bearing the sexual organs. PTERIDĂHYTA. A group of flowerless plants embracing ferns, horsetails, club mosses, etc. PUBESCENT. Covered with fine, soft hairs; downy. R ᴸHIS. The continuation of the stipe through the blade or leafy portion of the fern. REFLEXED. Bent abruptly downward or backward. RENIFORM. Kidney-shaped. REVOLUTE. Rolled backward from the margin or apex. ROOTSTOCK. (Or rhizome.) An underground stem, from which the fronds are produced. SCAPE. A naked stem rising from the ground. SEGMENT. One of the smaller divisions of a pinnatifid frond. SERRATE. Having the margin sharply cut into teeth pointing forward. SₒRULATE. The same only with smaller teeth. SESSILE. Without a stalk. SINUS. A cleft or rounded curve between two lobes. S̩UATE. With strongly wavy margins. SORUS A cluster of sporangia; a fruit dot. (plu. SORI). SP⁋ULATE. Shaped like a druggist's spatula or a flattened spoon. SPIKE. An elongated cluster of sessile sporangia. SP̩ULOSE. Spiny; set with small, sharp spines. SPORANGE (plu. A spore case. A tiny globe in which SPORANGIA). the spores are produced. STIPE. The stem of a fern from the ground up to the leafy portion; the leaf stalk. STOLON. An underground branch or runner. SᴢULATE. Awl-shaped. TₒNATE. With three nearly equal divisions. TRUNCATE. Ending abruptly as if cut off. TUFT. Things flexible, closely grouped into a bunch or cluster. VENATION. The veining of a frond or leaf. VERNATION. The arrangement of leaves in the bud. WHORL. A circle of leaves around a stem. WINGED. Margined by a thin expansion of the rachis.

NOTE

The student should have some idea of the terms genus, species and variety, although they are not capable of exact definition.

A species, or kind, is in botany the unit of classification. It embraces all such individuals as may have originated in a common stock. Such individuals bear an essential resemblance to each other, as well as to their common parent in all their parts. E.g., the Cinnamon fern is a kind or species of fern with the fronds evidently of one kind, and of a common origin, and all producing individuals of their own kind by their spores or rootstocks. When such individuals differ perceptibly from the type in the shape of the pinn乄or the cutting of the fronds, we have varieties as frond •, inc乑m, etc. Or if the difference is less striking the word form is used instead of variety, but in any given case opinions may differ in respect to the more fitting term.

A genus is an assemblage of species closely related to each other, and having more points of resemblance than of difference; e.g., the royal fern, the cinnamon fern, and the interrupted fern are alike in having similar spore cases borne in a somewhat similar manner on the fronds, and forming the genus Osmunda. In like manner certain members of the clover group--red, white, yellow, etc., make up the genus Trifolium.

Thus individuals are grouped into species and species are associated into genera, and the two groups are united to give each fern or plant its true name, the generic name being qualified by that of the species; as in the cinnamon fern Osmnda (genus), cinnam • (species).

In the following list the first name is usually the one adopted in the text, and those that follow are synonyms.

Names printed in small capitals are those of the newer nomenclature, now adopted at the Gray Herbarium but not in the Manual.

ADIANTUM L. 1. Adiantum Capillus-Veneris L. 2. Adiantum pedatum L. Var. ALEUTICUM RUPR. ASPIDIUM SW. 3. Aspidium Boottii. Tuckerm. Dryopteris Boottii. (Tuckerm.) Underw. THELYPTERIS BOOTTII. (Tuckerm.) Nieuwl. 4. Aspidium cristatum. (L.) Sw. Dryopteris cristata. (L.) A. Gray. THELYPTERIS CRISTATA. (L.) Nieuwl. 5. Aspidium cristatum var. Clintonianum. D.C. Eaton. Dryopteris cristata var. Clintoniana. (D.C. Eaton.) Underw. THELYPTERIS CRISTATA var. CLINTONIANA. (D.C. Eaton.) Weatherby. 6. Aspidium cristatum ɔmarginale. Davenp. 7. Aspidium Filix-mas. (L.) Sw. Dryopteris Filix-mas. (L.) Sw. THELYPTERIS FILIX-MAS. (L.) Nieuwl. 8. Aspidium fragrans. (L.) Sw. Dryopteris fragrans. (L.) Schott. THELYPTERIS FRAGRANS. (L.) Nieuwl. 9. Aspidium Goldianum. Hook. Dryopteris Goldiana. (Hook.) A. Gray. THELYPTERIS GOLDIANA. (Hook.) Nieuwl. 10. Aspidium marginale. (L.) Sw. Dryopteris marginalis. (L.) A. Gray. THELYPTERIS MARGINALIS. (L.) Nieuwl. 11. Aspidium noveboracense. (L.) Sw. Dryopteris noveboracensis. (L.) A. Gray. THELYPTERIS NOVEBORACENSIS. (L.) Nieuwl. 12. Aspidium simulatum. Davenp. Dryopteris simulata. Davenp. THELYPTERIS SIMULATA. (Davenp.) Nieuwl. 13. Aspidium spinulosum. (O.F. Muell.) Sw. Dryopteris spinulosa. (O.F. Muell.) Kuntze. THELYPTERIS SPINULOSA. (O.F. Muell.) Nieuwl. 14. Aspidium spinulosum var. intermedium. (Muhl.) D.C. Eaton. Dryopteris spinulosa var. intermedia. (Muhl.) Underw. THELYPTERIS SPINULOSA var. INTERMEDIA. (Muhl.) Nieuwl. 15. Aspidium spinulosum var. concordianum. (Davenp.) Eastman. THELYPTERIS SPINULOSA var. CONCORDIANA. (Davenp.) Weatherby. 16. Aspidium spinulosum var. dilatatum. (Hoff.) Gray. Dryopteris spinulosa var. dilatata. (Hoff.) Underw. THELYPTERIS SPINULOSA var. AMERICANA. (Fisch.) Weatherby. 17. Aspidium thelypteris. (L.) Sw. Dryopteris thelypteris. (L.) A. Gray. THELYPTERIS PALUSTRIS. Schott. ASPLENIUM L. 18. Asplenium Bradleyi. D.C. Eaton. 19. Asple-

nium platyneuron. (L.) Oakes. Asplenium ebeneum. Ait. 20. Asplenium ebenoides. R.R. Scott. 21. Asplenium montanum. Willd. 22. Asplenium parvulum. Mart, and Gal. Asplenium resiliens. Kze. 23. Asplenium pinnatifidum. Nutt. 24. Asplenium Ruta-muraria. L. 25. Asplenium Trichomanes. L. 26. Asplenium viride. Huds. ATHYRIUM. ROTH 27. ATHYRIUM ACROSTICH-OIDES. (Sw.) Diels. Asplenium acrostichoides. Sw. Asplenium thelypteroides. Michx. 28. ATHYRIUM ANGUSTIFOLIUM. (Michx.) Milde. Asplenium angustifolium. Michx. Asplenium pycnocarpon. Spreng. 29. ATHYRIUM ANGUSTUM. (Willd.) Presl. Athyrium filix-femina. American Authors not Roth. Asplenium filix-femina. American Authors not Bernh. 30. ATH-YRIUM ASPLENIOIDES. (Michx.) Desv. BOTRYCHIUM. SW. 31. Botrychium lanceolatum. (Gmel.) Angstroem. BOTRYCHI-UM ANGUSTISEGMENTUM. (Pease and Moore.) Fernald. 32. BOTRYCHIUM DISSECTUM. Spreng. Botrychium obliquum var. dissectum. (Spreng.) Clute. 33. Botrychium obliquum. Muhl. BOTRYCHIUM DISSECTUM var. OBLIQUUM. (Muhl.) Clute. 34. Botrychium lunaria. (L.) Sw. 35. Botrychium ramosum. (Roth.) Aschers. Botrychium matricariæfolium. A. Br. Botrychium neglectum. Wood. 36. Botrychium simplex. E. Hitchcock. 37. Botrychium ternatum. (Thunb.) Sw. Var. intermedium. D.C. Eaton. Botrychium obliquum var. intermedium. (D.C. Eaton.) Underw. 38. Botrychium virginianum. (L.) Sw. CAMPTOSO-RUS. LINK 39. Camptosorus rhizophyllus. (L.) Link. CHEILAN-THES. SW. 40. Cheilanthes alabamensis. (Buckley.) Kunze. 41. Cheilanthes Feei. Moore. Cheilanthes lanuginosa. Nutt. 42. Cheilanthes lanosa. (Michx.) Watt. Cheilanthes vestita. Sw. 43. Cheilanthes tomentosa. Link. CRYPTOGRAMMA.R. BR. 44. Cryptogramma densa. (Brack.) Diels. Pellæa densa. (Brack.) Hook. 45. Cryptogramma Stelleri. (Gmel.) Prantl. Pellæa gracilis. (Michx.) Hook. 46. Cryptogramma acrostichoides. R. Br. CYS-TOPTERIS. BERNH. 47. Cystopteris bulbifera. (L.) Bernh. Filix bulbifera. (L.) Underw. 48. Cystopteris fragilis. (L.) Bernh. Filix fragilis. (L.) Underw. DENNSTÆTIA L'HER. 49. DENNSTÆTIA PUNCTILOBULA. (Michx.) Moore. Dicksonia pilosiuscula. Willd. LYGODIUM SW. 50. Lygodium palmatum. (Bernh.) Sw. NOTHOLÆA.R. BR. 51. Notholæna dealbata. (Pursh.) Kunze. Notholæna nivea var. dealbata. (Pursh.) Davenp. ONOCLEA L. 52. Onoclea sensibilis. L. 53. Onoclea Struthiopteris. (L.) Hoff. Struthiopteris Germanica. Willd. Matteuccia Struthiopteris. (L.) Todaro. PTERETIS NODULOSA. (Michx.) Nieuwl. OPHIO-

GLOSSUM. (TOURN.) L. 54. Ophioglossum vulgatum. L. Ophioglossum vulgatum var. minus. Moore. 55. Ophioglossum Engelmanni. Prantl. OSMUNDA.L. 56. Osmunda cinnamomea. L. 57. Osmunda Claytoniana. L. 58. Osmunda regalis. L. OSMUNDA REGALIS var. SPECTABILIS. (Willd.) Gray. PELLƁ. LINK 59. Pell氚atropurpurea. (L.) Link. 60. Pell氚glabella. Mett. PHEGOPTERIS FΛ 61. Phegopteris Dryopteris. (L.) F饮 THELYPTERIS DRYOPTERIS. (L.) Slosson. 62. Phegopteris hexagonoptera. (Michx.) F饮 THELYPTERIS HEXAGONOPTERA. (Michx.) Weatherby. 63. Phegopteris polypodioides F饮 THELYPTERIS PHEGOPTERIS. (L.) Slosson. Phegopteris Phegopteris. (L.) Underw. 64. Phegopteris Robertiana. (Hoff.) A. Br. Phegopteris calcarea. F饮 THELYPTERIS ROBERTIANA. (Hoff.) Slosson. POLYPODIUM.L. 65. Polypodium vulgare. L. 66. Polypodium polypodioides. (L.) Watt. Polypodium incanum. Sw. POLYSTICHUM. ROTH 67. Polystichum acrostichoides. (Michx.) Schott. Aspidium acrostichoides. Sw. Dryopteris acrostichoides. (Michx.) Kuntze. 68. Polystichum Braunii. (Spenner.) F饮 Dryopteris Braunii. (Spenner.) Underw. Aspidium aculeatum var. Braunii. Doel. 69. Polystichum Lonchitis. (L.) Roth. Aspidium Lonchitis. Sw. Dryopteris Lonchitis. Kuntze. PTERIS.L. 70. Pteris aquilina. L. Pteridium aquilinum. (L.) Kuhn. PTERIDIUM LATIUSCULUM. (Desv.) Maxon. PTERIDIUM LATIUSCULUM var. PSEUDOCAUDATUM. (Clute.) Maxon. SCHIZƁ.J.E. SMITH 71. Schiz氚pusilla. Pursh. 72. Scolopendrium vulgare. J.E. Smith. PHYLLITIS SCOLOPENDRIUM. (L.) Newman. TRICHOMANES.L. 73. Trichomanes radicans. Sw. Trichomanes Boschianum. Sturm. WOODSIA.R. BY. 74. Woodsia glabella. R. Br. 75. Woodsia alpina. (Bolton.) S.F. Gray. Woodsia hyperborea. R. Br. 76. Woodsia ilvensis. (L.) R. Br. 77. Woodsia Cathcartiana. B.L. Robinson. 78. Woodsia obtusa. (Spreng.) Torr. 79. Woodsia oregana. D.C. Eaton. 80. Woodsia scopulina. D.C. Eaton. WOODWARDIA.J.E. SMITH 81. Woodwardia virginica. Sm. 82. Woodwardia areolata. (L.) Moore. Woodwardia angustifolia. Sm. THE PETRIFIED FERN In a valley, centuries ago, Grew a little fern-leaf green and slender, Veining delicate and fibers tender, Waving when the wind crept down so low; Rushes tall and moss and grass grew round it, Playful sunbeams darted in and found it, Drops of dew stole down by night and crowned it. But no foot of man e'er came that way-- Earth was young and keeping holiday. Monster fishes swam the silent main, Stately forests waved their giant branches, Mountains hurled their snowy avalanches, Mammoth creatures stalked across the plain, Nature reveled in grand mysteries; But the little fern was not of these, Did not slumber with the hills and trees, Only grew and waved its wild, sweet way; No one came to note it day by day. Earth, one time, put on a frolic mood, Heaved the rocks and changed the mighty motion Of the deep, strong currents of the ocean; Moved the plain and shook the haughty wood, Crushed the little fern in soft, moist clay, Covered it and hid it safe away. Oh, the long, long centuries since that day! Oh, the changes! Oh, life's bitter cost! Since the useless little fern was lost. Useless? Lost? There came a thoughtful man Searching Nature's secrets far and deep; From a fissure in a rocky steep He withdrew a stone o'er which there ran Fairy pencilings, a quaint design, Leafage, veining, fibers clear and fine, And the fern's life lay in every line! So, I think, God hides some souls away, Sweetly to surprise us the last day!--M.B. BRANCH.